A Study of Teaching
Multiple Lenses, Multiple Views

A Study of Teaching
Multiple Lenses,
Multiple Views

Alan H. Schoenfeld, *Volume Editor*
University of California at Berkeley

Neil Pateman, *Series Editor*
University of Hawaii

NATIONAL COUNCIL OF
TEACHERS OF MATHEMATICS

Copyright © 2008 by
THE NATIONAL COUNCIL OF TEACHERS OF MATHEMATICS, INC.
1906 Association Drive, Reston, VA 20191-1502
(703) 620-9840; (800) 235-7566; www.nctm.org

Library of Congress Cataloging-in-Publication Data

Schoenfeld, Alan H.
 A study of teaching : multiple lenses, multiple views / Alan H. Schoenfeld, Neil
Pateman.
    p. cm. — (JRME monograph ; #14)
  ISBN 978-0-87353-603-5
  1. Mathematics—Study and teaching—Research. 2. Teaching—Research.
I. Pateman, Neil A. II. Title.
QA11.2.S36 2008
510.71—dc22
                            2007043266

The National Council of Teachers of Mathematics is a public voice of mathematics education, providing vision, leadership, and professional development to support teachers in ensuring equitable mathematics learning of the highest quality for all students.

Printed in the United States of America

Table of Contents

Contents of the CD

Video of Lesson Segment, Deborah Ball's Third-Grade Class, Friday, January 19, 1990

Transcript of Lesson Segment

Excel Spreadsheet, Chapter 2: Ball Lesson Transcript, Parsing, and Teacher Goals

Acknowledgments

This work has been supported by grants from the National Science Foundation (REC # 0126237) and the Spencer Foundation (MG # 199800202). The opinions expressed do not necessarily reflect the views of either funding organization.

The contributors to this monograph include Deborah Loewenberg Ball, Ilana Seidel Horn, Elham Kazemi, Jennifer Lewis, Tamar Posner, Alan H. Schoenfeld, Miriam Gamoran Sherin, Bruce L. Sherin, and Mark Hoover Thames.

Through the Looking Glass:
A Study of Teaching

Jennifer Lewis
University of Michigan

> Let us recognize that ... what has been said is not still to be said; that an expression does not have the same value twice, does not live two lives; that words, once spoken, are dead and function only at the moment when they are uttered, that a form, once it has served, cannot be used again and asks only to be replaced by another, and that the theater is the only place in the world where a gesture, once made, can never be made the same way twice. (Artaud, 1958, p. 75)

In the passage above from *The Theater and Its Double,* Antonin Artaud argues that only theater can convey the fleeting nature of experience. Teaching work, like theater, unfolds in real time, in the presence of others. No teaching moment is quite like the next; even if the teacher were to repeat the same script, its reception would vary with every delivery. This phenomenon is a central challenge in teaching as well as in studying teaching. Because its enactment flies by, and because each lesson or unit or encounter is not exactly like any other, teaching is difficult to describe even simply.

This volume represents the efforts of a number of scholars to study a single, brief excerpt of teaching in fine detail. A number of years ago, Alan Schoenfeld and the Teacher Modeling Group at the University of California—Berkeley invited Deborah Ball and the Mathematics Teaching and Learning to Teach Group at the University of Michigan to write separate analyses of a single, 6-minute video clip excerpted from a mathematics lesson in a third-grade classroom. Schoenfeld's invitation to have different individuals write about the same videotape was the spark that grew first into a conference symposium and finally into this monograph.

As varied as the chapters in this volume are, beneath the surface one detects shared points of departure. The first is that all the authors are interested in the *fine-grained analysis and study of teaching*. A second is that the authors intentionally solicit *multiple perspectives* around a common artifact of teaching. In this Introduction we describe those common starting points and give an overview of the four chapters that follow, each of which presents a different analysis of 6 minutes of a mathematics lesson in a third-grade classroom. We also consider what the four assembled chapters have to offer and the kind of knowledge they generate.

THE STUDY OF TEACHING

Why study teaching? Why devote so much analysis to an activity as prosaic as teaching?

Although relatively young as a field of study, research on teaching now has a history long enough to be punctuated by trends and chronological periods. The first *Handbook of Research on Teaching* (Gage, 1963) was published in 1963. By the publication of the *Second Handbook of Research on Teaching* (Travers, 1973), the quantity of research on teaching that had been conducted prompted Barak Rosenshine and Norma Furst to note in that volume that, as a body, it was "chaotic," "unorganized," and a "maze of instrumentation and research" (p. 122). By the time the *Handbook of Research on Teaching* was published in 1986, Lee Shulman found it necessary to develop a typology for sorting the many different types of research on teaching by paradigms and research programs. There, Shulman details the variation and complexity of classroom life and various ways to study it. He points out that the unit of analysis in research on teaching can vary: "Is it the individual interchange between student(s) and teacher, the episode (e.g., quelling a particular behavioral disturbance, or explaining a new concept), the lesson (say, a 20-minute reading group session), the unit (e.g., a six-day sequence on the Age of Jackson in a U.S. history course), the semester course, or the year of work?" (p. 8) He goes on to comment that researchers can vary in their conceptions of content (topical, substantive vs. syntactic, etc.), in their research perspectives (positivist, interpretive, eclectic, etc.), and in their methods (natural observation, laboratory studies of cause-and-effect, surveys of teacher beliefs, etc.). This variation continues to be reflected in the *Handbook of Research on Teaching* (Richardson, 2001), which reports flourishing research programs of all shades and stripes, yet reveals that the field of research on teaching remains undeveloped in comparison with, say, the field of research on learning.

Despite the multiple approaches to research on teaching and the many studies conducted over the years, this research has yielded surprisingly little about the basic conduct of classroom life and what teachers do so that children will learn. Consider the example of "classroom discourse." In plain language, researchers probe how teachers and children talk in the classroom. We can think of studies of classroom discourse in a wide range of research programs that together constitute a huge body of scholarly knowledge about talk in schools. We know from Shirley Brice Heath's work (1983), for example, how socioeconomic and ethnic affiliations shape language use at home and how this influence plays out in school settings. Courtney Cazden's studies in elementary school classrooms (2001) reveal patterns in how teachers typically talk with children. In mathematics classrooms specifically, Edwards and Mercer (1987), Setati, Adler, Reed, and Bapoo (2002), and Kieran, Forman, and Sfard (2003) have all contributed to our understanding of classroom discourse. We think we can safely say that a strong corpus of knowledge now exists regarding classroom talk and even classroom talk in mathematics. But the work of constructing a productive discussion in a mathematics class—crafting an opening question, soliciting wide responses, asking generative probes, responding differentially to student productions, drawing toward closure or steering clear of it—these pedagogical moves are the bread and butter of teaching work, and yet research on teaching has offered little that is directly, or even indirectly, usable by teachers in the minute-by-minute, experience-based daily work they do. The studies in this volume add to a growing attempt to produce such knowledge. They join such lines of research as those by Lampert (2001) and O'Connor and Michaels (1996).

The fact that teaching work remains largely underspecified may seem surprising. Perhaps the very nature of teaching as a commonplace activity makes it difficult to analyze. Although maladroit teaching is often criticized, its opposite, skillful teaching, is often attributed to talent or instinct rather than accomplished technique. In fact, the word *technique* is anathema to many, who prefer to think of teaching as art, neglecting to consider that artistic virtuosity also involves substantial command of technique. As Margolis has written, "When everyone in a community shares a habit, it ordinarily becomes invisible, for what everyone does no one easily recognizes" (1993, p. 17). Teaching is such a "habit," one that is not only ordinary to its practitioners but perhaps, unlike other professions, ordinary to nearly everyone. Nearly all adults know the work of teaching firsthand from their many years of schooling as children, during what Lortie so elegantly labeled their "apprenticeship of observation" (1975, p. 61). On the one hand, then, the work of teaching is everyday and therefore evades notice. On the other hand, teaching that does not conform to traditional images is hard to understand and also resists analysis. The implication is that teaching is therefore difficult to understand or to learn to do. Lacking a well-articulated analysis, teachers are hampered in learning essential features of the work beyond mimicking the memory of what they think was done to them as children. The profession lacks a rich and nuanced shared language with which to discuss and improve the work in the company of colleagues.

So not for want of rigorous, valuable research on teaching do we engage in the study of teaching contained in this monograph. The studies here draw heavily from features of past and current research on teaching. For example, each of the four chapters makes heavy use of discourse analysis. Schoenfeld's chapter focuses on the role of teacher beliefs and knowledge. The chapter by Horn and the chapter by Posner give particular emphasis to children's access to, and participation in, classroom activity. All chapters treat cognition without labeling it as such. All chapters devote attention to the subject matter knowledge of the teacher and make claims about children's learning. Although process-product research in the form of the classic descriptive-correlational-experimental loop seems to have fallen from favor, all four chapters in this volume consider, implicitly or explicitly, how teacher behaviors in this 6-minute videotape are linked with student outcomes even if those outcomes are not expressed in standardized test scores. Thus, we draw on the foundational work of many researchers, and hope that the contents of this volume will be seen as a refinement and extension of prior research.

At the same time, this work has a distinctly pragmatic focus. The studies of teaching in this monograph are intended to contribute to a growing effort to pursue research on teaching that can influence the work that teachers do in classrooms. In doing so, they join a wider oeuvre of analysis of everyday work in other professions. To take an example that at first seems far afield, consider Mike Rose's study of his mother's work as a waitress (2001). His analysis drives home the notion that even such "unskilled" labor as waitressing involves serious cognitive and physical complexity. "The layout of the tables (or the stools at the counter) and people's location at them enabled my mother to store and recall information about orders in a number of ways. A customer's specific position (by the window or closest to her) mattered, especially if it were somehow unusual—for example, a woman pulls a fifth chair to the edge of a four-top. Relative locations also figure in, aided by other characteristics of the person or the order" (p. 12).

Rose wonders about the cognitive processes that waitresses use in the face of conflicting and multiple demands. He writes in his analysis of his mother's work that "there is no experimental cognitive science literature on this dynamic mix of attending and immediate decision making that maps directly onto waitressing. This dimension of the work would be, I imagine hard to simulate and study in the laboratory with any marked ecological validity" (p. 14). Although teaching has been studied far more than waitressing, both professions, and others, share a paucity of analyses that capture the integrated complexity of multifaceted work.

An exception is nursing. Multiple studies of nursing work have revealed elements of the practice that are little articulated even by expert practitioners or observers of nursing practice. Bowker and Star (1999) refer to the Nursing Interventions Classification developed at the University of Iowa, which includes such details of nursing practice as the use of humor in dealing with patients. The manual page on humor includes the following definition: "Facilitating the patient to perceive, appreciate, and express what is funny, amusing, or ludicrous in order to establish relationships, relieve tension, release anger, facilitate learning, or cope with painful feelings" (p. 21). Benner (2001) convened groups of expert nurses who articulated the following domains of practice:

The Helping Role

The Teaching-Coaching Function

The Diagnostic and Patient-Monitoring Function

Effective Management of Rapidly Changing Situations

Administering and Monitoring Therapeutic Interventions and Regimens

Monitoring and Ensuring the Quality of Health Care Practices

Organizational and Work-Role Competencies (p. 46)

We are far from such an articulated parsing of the knowledge and skill domains for teaching, and we believe that such shared knowledge would help novices learn to teach and would provide a professional language that would help experienced teachers continue to improve practice. The chapters in this volume contribute to such articulations by attempting to analyze, label, and understand what teachers do in practice.

THE CALL FOR MULTIPLE LENSES

The world of classroom life is actually many simultaneous worlds, lying side by side or nested one within another, or completely removed one from another. The chapters in this monograph take the conversation between the teacher and students as the story line of the 6 minutes we see in the videotape, but in truth many story lines are going on simultaneously. Some children might not be listening to that discussion, or other children might be talking among themselves. What transpires in those 6 minutes is part of a much bigger story, located as it is within a school, a school district, a community, a country. One could place the 6-minute segment against the backdrop of educational policy, or analyze how the teacher and children play out socioeconomic structures of which school is a part. The "real" classroom experience is elusive: each moment is experienced differently by the actors involved and their perceptions of those experiences change with time and reflection. The choices of what to focus on, which story to follow, are endless.

Analyzing classroom work, then, bedevils researchers: Quite simply, no definitive version of what "happened" exists, nor can definitive meanings be ascribed to what happened. Often, even to *see* "what happened" can be difficult. As the chapters in this volume attest, even what is seen through a single camera lens for only 6 minutes in one classroom produces wide variations in interpretation. Because teaching is a complex activity, and because it involves multiple actors and "standpoints" (Harding, 1993), we are committed to the intentional gathering of such diverse interpretations of classroom teaching. In his essay "The Practical: Arts of Eclectic" (1978), Joseph Schwab describes how individual theories from social science are inadequate in addressing problems of practice, but indicates that bringing multiple theories to bear on such problems illumines them so that progress can be made.

Schwab notes that problems of practice "are never solved completely or once and for all" (p. 322), and he attributes this phenomenon in part to the disparities between theory and practice. Although practical problems in teaching are situation-specific—particular to time, place, person, and circumstance—theory leaves behind particulars of abstraction and generality. Schwab proposes that theory can be readied for practical use through "the arts of eclectic" (p. 323) such that the distortions and limited perspectives that theory imposes can be discovered and counterbalanced. Thus, across the chapters in this monograph we find multiple theoretical perspectives, even though these perspectives may present conflicting or inconsistent views. We intentionally seek alternative and competing perspectives on problems of practice, for the reason that no one theory can sufficiently illumine what is by nature a complex object of study. In the chapters here we see discourse theory, the construct of "situated cognition," and sociological theory—just to name a few—used to understand the passage of talk seen in this classroom excerpt. These theories are neither entirely compatible nor in all instances complementary, but each provides a useful if competing lens that produces varied understandings of the passage.

Take, for example, an exchange between Sheena and Sean early in our 6-minute classroom excerpt, turns 1–20 in the transcript found in Appendix 1 to this volume.

Ball:	And I have a few questions. First my first question is, I'd just like to hear some comments about what you thought about the meeting, what you noticed about the meeting, what you learned at the meeting, just what kinds of comments you have about yesterday's meeting? And could you listen to one another's comments, so that we can, um, benefit from what other people say? See what y— what you think about other people's comments? Sheena, do you want to start?
Sheena:	I—I—I liked it because, well, I like talking to other classes and, and when you talk to other classes sometimes it helps.
Ball:	In what way?
Sheena:	It helps you to understand a little bit more.
Ball:	Was there an example of something yesterday that you understood a little bit more during the meeting?
Sheena:	Well, I didn't think that zero was—zero, um—even or odd until yesterday they said that it could be even because of the ones on each side is odd, so that couldn't be odd. So that helped me understand it.
Ball:	Hmm. So y— So you thought about something that came up in the meeting that you hadn't thought about before? Okay.
Sheena:	*(nods)*
Ball:	Other people's comments? Sean?
Sean:	Um, I—I—I just want to say something to Sheena, when sh— what she said about um that, that one, um—zero has to be an odd, an even number bec— I disagree because, um, because what what two things can you put together to make it?
Sheena:	Could you repeat what you said, please?
Ball:	*(speaks to Betsy and asks her to listen to this)*
Sean:	Okay, um, I disagree with you because, um, if it was an even number, how— what two things could make it?
Sheena:	Well, I could show you it. *(Moves toward the chalkboard and points to the number line above the chalkboard.)* Um, I forgot what his name was—but yesterday he said that this one *(points to the 1 on the number line)* and each—this one is odd and this one *(points to the –1 on the number line)* is odd, so this one has to be even.
Sean:	But, that doesn't mean it always is even.
Sheena:	It could be even.

Sean:	It could be, but . . .
Sheena:	I'm not saying that is has to be even. I meant that it could be.
Sean:	You said it was.
Ball:	Before we take this up again, I underst— I— I understand that this is still a problem and that we didn't a— we didn't settle it, we're probably not going to settle it. Um, there's a lot of disagreement about this issue, right? And you saw that the fourth graders who have been thinking about this for a long time also disagree about it, don't they?

The reader will find that the four chapters in this volume offer conflicting interpretations of this exchange. Ball, Lewis, and Thames frame this "breakdown" between Sheena and Sean as a clash of two seemingly irreconcilable definitions for *even* and *odd*. Schoenfeld characterizes this passage of talk as a "dilemma" for the teacher along three dimensions: the diversion from the teacher's stated agenda, the differing mathematical views between students, and the "tension" around "issues of ego and identity at stake for both students." Posner reads this same short episode and suggests that Sheena "withdraws" from the exchange entirely. She attributes this withdrawal in part to the teacher's inattention to issues of equity at play in that moment. Horn interprets these turns differently still: she holds that the teacher "supports" Sheena, that Sean "redirects" the conversation from a reflective one to an argumentative one, and that both students experience this spate of turns with "relational discomfort."

In our view, none of these interpretations is singularly correct. Taken together, they do not rest easily. How could they, when some interpretations contradict one another entirely? The assembly of a multiplicity of viewpoints increases our ability to see more in this exchange. When using equity as a prism, for example, the words and actions in the classroom are refracted anew. The aim is neither to attain theoretical purity nor to illustrate theory with concrete examples from classroom work. Rather, these competing interpretations are arrayed here so that we can better understand and explicate practical problems in teaching. With more to see, teachers have more ways to act.

THE USE OF NEW TECHNOLOGIES

We return for a moment to the passage by Artaud (1958) quoted at the beginning of this Introduction. In real time, in lived experience, words will never be repeated in just the same way as they were once. Gestures vary despite verbatim repetitions of talk. All of experience is fleeting and ephemeral, and teaching is no exception. Videotape is seen as one vehicle that suspends the action and holds it still long enough for it to be examined. This medium has made possible a certain kind of iterative analysis, in that one can play and replay the moment of classroom action captured by the camera. And the existence of videotape plays a significant role enabling analyses to be shared across authors. We know that videotape can be a powerful tool for research (see, e.g., Granott Farber, 1990). But in this particular endeavor, videotape made possible researchers' independent look at an episode of classroom teaching in spite of separation by time and geographic distance. Videotape does not provide an objective rendering of events, nor does it contain the "whole picture," for we see only what the videographer chose to capture. Yet videotape is less processed and more authentic to the events than other forms of data to which scholars of teaching usually have access. Compared with field notes, interviews, audiotape, written transcripts, and other forms of data that researchers typically use, videotape is closer to the events as they happened with less processing by an intermediary. We see what Mei actually said in class and hear her inflections rather than rely on our memory of her comment or an ethnographic account of it. This feature was especially useful in working across groups whose members bring different perspectives and methods to the analysis of teaching.

WHAT KIND OF KNOWLEDGE IS BEING PRODUCED HERE?

The studies in this monograph produce at least three different kinds of knowledge. One kind is about the quotidian practices of keeping mathematics teaching going. The multiple analyses—from different standpoints, about different "realities" in the classroom, using different lenses focused on different worlds of classroom life—render a textured portrait of teaching. The varied presentations in this volume help us "see" more of what is to be seen in teaching. We think this knowledge is usable by practitioners and researchers.

Second, the work across chapters helps each author see more clearly his or her own frame of reference. Not only the object of study, the videotape segment, is brought into focus through these contrasting portraits of classroom life; the very lenses themselves also come to be more explicit when positioned one next to the other. Implicit assumptions are prodded to the surface, routine perceptions are jarred. Thus, the work in this volume helps us understand both teaching and the tools we use for researching teaching.

Third, as a set, these chapters strike an epistemological stance, perhaps unintentionally. Knowing teaching entails seeing from multiple angles. One cannot know teaching fully; a monograph of many more chapters would not uncover all that can be seen in the videotaped 6 minutes of classroom life. We return to the discussion of Sheena and Sean as an example. The assembly of multiple analyses is not offered so that one can trump the others. We do not present these analyses in the belief that one of them approaches the "proper" interpretation of what transpired in this classroom segment, and neither do we claim that through assembling them we are closer to some single correct version of what happened in those 6 minutes.

This stance prompts a number of questions, however. How does one makes judgments about the quality or validity of each analysis of the 6 minutes, illustrated most pointedly by the number of instances in which the chapters contradict one another in interpreting just what happened in those 6 minutes? Consider the broader question of how disagreement is managed in this 6-minute excerpt from the classroom. In the chapter by Ball, Lewis, and Thames, the authors make the following claim: "Notable across these and the other examples is how entirely comfortable the children seem to be about receiving public disagreement, and how attentively they listen to it and respond. Despite the fact that they are engaged in conversations that involve substantial disagreement and agreement, students' tone and manner are strikingly dispassionate and matter-of-fact" (chapter 1, p. 22).

The other authors of chapters in this monograph interpret disagreement quite differently. Horn implies that social discomfort threatens, and that what she calls "accountable argumentation" mediates this discomfort: "Accountable argumentation, with its expectation of modifying positions in light of convincing evidence, provides an interactional resource for Sean to 'revise' his position without losing face" (chapter 3, p. 120). Posner sees disagreements as shifting from the conceptual to the personal, so offers the following advice: "Through our intervening and keeping the conversation focused on concepts, classrooms might become safer spaces for students who traditionally do not speak out in discussions. Thus, identifying and intervening in such moments can become a practical way to address equity" (chapter 4, p. 161). Schoenfeld conceives of disagreement in this classroom as an important discourse practice introduced by the teacher to support mathematical learning. "Norms [established by the teacher] are evident for interacting on an appropriately respectful basis (for example, students say, 'I disagree,' not 'You're wrong')" (chapter 2, p. 62).

The foregoing examples showing differing views about disagreement in the classroom prompt us to wonder: When multiple perspectives are sought, what is meant by "get it right" in this kind of work? Is the analysis therefore valid—"valid" here in the sense of "well founded and fully applicable to the particular matter or circumstances; sound and to the point; against which no objection can fairly be brought" (*Oxford English Dictionary, 2nd Edition,* 1989)? What are the criteria for judging the usefulness, the strengths, the rightness of each analysis when one is committed to an assembly of necessarily competing views? Our conjecture is that an analysis is useful if it helps the reader "see" more in the classroom.

WHAT THIS MONOGRAPH CONTAINS

Following this Introduction is a brief synopsis of the 6-minute fragment of a third-grade mathematics lesson that all authors analyzed. Next is a brief description of the database from which the videotape was taken. Then follow four chapters, each analyzing that 6-minute videotape. The first chapter, by Ball, Lewis, and Thames, is primarily concerned with the work the teacher does to help students make mathematical claims. Next follows Schoenfeld's fine-grained analysis of teaching, in which he builds a model of the teacher's decision making, as shaped by her beliefs, goals, and knowledge. Horn's chapter uses the data from this 6-minute videotape to illustrate the participation structures in the classroom that help support productive disagreements. She demonstrates how accountable argumentation supports students in constructing mathematical understanding in a conversational medium that harnesses disagreement but mediates its social risks. Posner uses equity as a lens for understanding the 6-minute episode, and she attributes students' participation to functions of equity as they play out in this classroom. These four chapters are followed by two commentaries, one by Elham Kazemi and the other by Miriam Sherin and Bruce Sherin. A transcript of the 6-minute video clip is found in Appendix 1 to this volume.

A SYNOPSIS OF THE 6-MINUTE VIDEO SEGMENT

This lesson takes place on a Friday in mid-January. A class of third-grade students has just come in from a cold recess on the playground. The teacher, Mrs. Ball, begins class by asking for "comments" about yesterday's meeting with the fourth graders, a special event at which they had discussed even and odd numbers, and zero, with the older children. Sheena volunteers. She says she liked the meeting because "when you talk to other classes, it sometimes helps." When Mrs. Ball asks for an example, Sheena elaborates—she says she had not thought zero was even or odd. When she heard one of the fourth graders explain that it could be even "because of the ones on each side is odd," she had begun to think more. Her teacher nods, "So you thought about something that came up in the meeting that you hadn't thought about before? Okay." And the teacher asks for other people's comments.

Sean raises his hand, but rather than comment about the experience of the meeting, he says he has something he wants to say to Sheena about zero's being even. He turns to her. "I disagree, because, um, because what two things can you put together to make it?" She explains, using the number line, but Sean remains unconvinced—saying that no two things can be put together to make "it" (zero). Ball intervenes. "We're probably not going to settle [this]," she says, noting the impasse the two have reached for the moment. Again she asks for other comments about the meeting.

Across these first 6 minutes of class, students make claims, query one another, and revise their positions. The pace is quick. The teacher clarifies, presses for reasons, and orchestrates the discussion. For example, when Nathan says that he thinks zero "is special" and that even numbers "make even numbers," Mrs. Ball asks what he means, "Were you saying that when you put even numbers together, you get another even number—or were you saying that *all* even numbers are made up of even numbers?" "Yeah," Nathan replies, seeming to agree with both alternatives. The teacher asks Betsy to comment, reminding her, "You said something like that yesterday, too." As Betsy interacts with Nathan, other students chime in. They pick up on her ideas and add their own. "Like six is two . . . ," begins Betsy. Sean completes her sentence: "Six is two *odd* numbers to make an even." "Three and three," adds Mei. As the exchanges among the children continue, Nathan's claim becomes clearer—he is not talking about every even number.

When the teacher asks for more comments about the meeting, Sean says, "I don't have anything about the meeting yesterday, but I was just thinking about six … , that it could be an odd and an even number." Initially struck by an earlier statement of Betsy's that six is *three twos,* we can see now why it has occurred to him to make this claim—that six is two threes and therefore is even. But because it is also three twos, "and three of something is, like, *odd,*" six can also be considered odd, he argues.

ABOUT THE DATA SET

The data set from which this selection of videotape is drawn comprises records from an entire year of a third-grade class taught by Deborah Ball in 1989–1990. Ball taught mathematics daily in this public school classroom; her colleague, Magdalene Lampert, taught fifth grade daily at the same public school. A grant from the National Science Foundation to Ball and Lampert enabled them to document the year of classroom instruction. Classes were videotaped across the year, and all the students' work and the teacher's notebook were photocopied. Other artifacts—the students' homework, tests and quizzes, report cards, and the curriculum—were also preserved. Since the year of documentation, the researchers and their teams have explored ways to digitize the materials and create tools for accessing and investigating the two classrooms. Ball and Lampert documented their teaching in these ways, not to model "good" or "exemplary" teaching but to create a resource for the study of teaching, of which the work included in this volume is an example.

REFERENCES

Artaud, A. (1958). *The theater and its double.* New York: Grove Press.

Benner, P. (2001). *From novice to expert: Excellence and power in clinical nursing practice.* Upper Saddle River, NJ: Prentice Hall Health.

Bowker, G. C., and Star, S. L. (1999). *Sorting things out: Classification and its consequences.* Cambridge, MA: MIT Press.

Cazden, C. B. (2001). *Classroom discourse: The language of teaching and learning.* Portsmouth, NH: Heinemann.

Edwards, D., and Mercer, N. (1987). *Common knowledge: The development of understanding in the classroom.* London: Methuen.

Gage, N. S. (Ed.). (1963). *Handbook of research on teaching.* Chicago: Rand McNally.

Granott Farber, N. (1990). Video as a research tool. In *Constructionist learning* (pp. 319–326). Cambridge, MA: MIT Media Lab.

Kieran, C., Forman, E. A., & Sfard, A. (2003). *Learning discourse: Discursive approaches to research in mathematics education.* New York: Springer.

Lampert, M. (2001). *Teaching problems and the problems of teaching.* New Haven: Yale University Press.

Lortie, D. C. (1975). *Schoolteacher: A sociological study.* Chicago: University of Chicago Press.

Margolis, H. (1993). *Paradigms and barriers: How habits of mind govern scientific beliefs.* Chicago: University of Chicago Press.

O'Connor, M. C., and Michaels, S. (1996). Shifting participant frameworks: Orchestrating thinking practices in group discussion. In D. Hicks (Ed.), *Child discourse and social learning* (pp. 63–102) Cambridge: Cambridge University Press.

The Oxford English Dictionary (2nd ed). (1989). *OED Online.* Oxford University Press. Retrieved October 12, 2006, from http://ets.umdl.umich.edu/cgi/o/oed/oed-idx?q1=valid&type=Lookup

Richardson, V. (Ed.) (2001). *Handbook of research on teaching* (4th ed.). Washington, DC: American Educational Research Association.

Rose, M. (2001). The working life of a waitress. *Mind, Culture, and Activity, 8*(1), 3–27.

Schwab, J. J. (1978). The practical: Arts of eclectic. In Westbury, I., & Wilkof, N. J. (Eds.), *Science, curriculum, and liberal education* (pp. 322–364). Chicago: University of Chicago Press.

Setati, M., Adler, J., Reed, Y., & Bapoo, A. (2002). Incomplete journeys: Code-switching and other language practices in mathematics, science and English language classrooms in South Africa. *Language and Education, 16*(2), 128-149..

Shulman, L. S. (1986). Paradigms and research programs in the study of teaching: A contemporary perspective. In M. C. Wittrock (Ed.), *Handbook of research on teaching* (pp. 3–36). New York: Simon & Schuster Macmillan.

Travers, R. M. W. (Ed.). (1973). *Second handbook of research on teaching.* Chicago: Rand McNally.

Chapter 1

Making Mathematics Work in School[1,2]

Deborah Loewenberg Ball
Jennifer Lewis
Mark Hoover Thames
University of Michigan

We begin with a well-known classroom episode. It takes place on a midwinter day in third grade. About 6 minutes into mathematics class, a boy named Sean[3] raises his hand and says he has been "thinking about the number six." He says he has been thinking it could be even or it could be odd. His is a peculiar claim, because these third graders already know that six is an even number. They learned that fact in second grade. In the ensuing minutes, his classmates, sure that six cannot possibly be odd, challenge him and try to show him the fallacy in his thinking. Still, significantly, no one laughs at this exchange as mathematical foolishness. And they do not simply contradict him, nor refer to some "fact" learned somewhere else. Instead, they press him to "prove" his idea. Listening to them, one can detect that "prove it" constitutes a serious intellectual request, not a social taunt.

Sean persists. He has noticed something special about six that he finds interesting: it is made up of two threes, but also of three twos. Since two is even and three is odd, six seems to him to have both even *and* odd structures. Sean thinks this composition is interesting. Sean's idea agitates many of his classmates, who are quite sure six is not odd. They argue with him, and Tembe challenges Sean to "prove it" to the class. At this point in the class, the teacher presses the students to consider their working definition for even numbers. One of Sean's classmates, Mei, listening intently, understands. "I think I know what he is saying," she says. "I think what he's saying is that you have three groups of two. And three is an odd number so six can be an odd number *and* a even number." Sean is pleased: Mei has grasped his idea. However, then Mei says she disagrees with this reasoning. According to Mei, whether a number is considered even or odd is not according to how many groups of two it has. To provide an example, she draws 10 circles on the chalkboard (see Figure 1.1).

---

[1] This work has been supported by grants from the National Science Foundation (REC # 0126237) and the Spencer Foundation (MG #199800202).

[2] We gratefully acknowledge our colleagues Hyman Bass and Edward Wall for their influence on the thinking and work reflected in this article.

[3] All names are pseudonyms, standardized across published analyses of these data, and selected to be culturally similar to the children's real names. For example, Sheena, an African American child, was given a name chosen from among other moderately common African American girls' names; Tembe was from Kenya, and his pseudonym was selected from among similar Kenyan boys' names.

o o|o o|o o|o o o|o o

Figure 1.1. Mei's drawing of 10 circles as five groups of two.

Mei shows Sean that if he extended his observation of six to other numbers—in this example, ten—other numbers would be similarly "even and odd." Ten has five groups of two. She thinks he will see the error and retract his claim. Instead, Sean acknowledges that ten can be even and odd, too. This idea sends the class into pandemonium. Mei is amazed:

> What about *other* numbers?! Like, if you keep on going on like that and you say that other numbers are odd and even, maybe we'll end it up with *all* numbers are odd and even. Then it won't make sense that all numbers should be odd and even, because if all numbers were odd *and* even, we wouldn't be even having this *discussion!*

In the next 20 minutes, other students enter the fray, and eventually the class has realized that Sean's observation of six applies to other numbers as well: 14, 18, 2, 26, ..., that is, all the odd multiples of two. The teacher eventually labels these special evens as "Sean numbers," properly distinguished from the terms *even* and *odd*. The pupils explore the properties of these new numbers.

This remarkable episode, briefly summarized in the foregoing, has served as the subject of many analyses and much discussion, in the research literature and beyond. The episode is one excerpt from a single mathematics lesson, and video clips from this lesson have been used to stimulate broad conversation about education, mathematics, teaching and learning, teacher development, and policy. Less examined, however, is the backdrop of these pupils' and their teacher's work. Sean's idea was percolating for several minutes before he brought it up in class. He and his classmates had also already been practicing ways to explore mathematical ideas and develop conjectures about them. This article makes that backdrop the focus. What constitutes doing such mathematical work in school?

Lampert (2001) argues that learning and teaching mathematics in school are constructed interpersonally, socially, and intellectually across time.

> The different units of time in which teaching happens are one source of its complexity. Teaching problems are solved in particular moments of interaction, and in the larger scale of the lesson as a whole. . . . The teacher also acts across groups of lessons, ranging from a pair of lessons connected across two days to the totality of all lessons across the year. Still other actions, tasks, or strategies may need to be performed to maintain longer-term connections between students and mathematics. The complexity of teaching in time does not end with considering different units, however, because teaching acts are not uniquely situated in single units of time. (p. 36)

To gain perspective on this oft-discussed episode, we examine the 6 minutes of class that lead up to Sean's announcement that six "could be an odd and an even number, both."[4] We use this segment to explore the nature and demands of doing mathematical work of the kind in which Sean and his classmates were engaged. Our purpose is to unpack the work of teaching in ways that make it accessible for scrutiny, analysis, and development. We argue that a central task of teaching is making mathematical work for students, that intellectually honest mathematical work can be done in school, and that much of the work of teaching is itself mathematical work.

MATHEMATICAL WORK

Teacher: Um, there's a lot of disagreement about this issue, right? And you saw that the fourth graders who have been thinking about this for a long time also disagree about it, don't they? (turns 20B–20C)[5]

Is zero even or odd? Or is it "special"? What is the answer to 0 minus 9? Is 4/4 smaller or larger than 5/5, or are those quantities the "same"? Questions such as these fill the days of elementary school teachers as students wonder and puzzle about mathematics. Even if many adults can answer these questions for themselves, how eight-year-olds might sensibly reason about them is not immediately obvious.

The challenge for the elementary-grades teacher grows from the imperative that answers be based on mathematical reason in teaching and learning mathematics. After all, zero is even because the definition of an even number is an integer multiple of two, and zero is 0 times 2. But even if eight-year-olds *have* a definition, it does not come in this form: In fact they likely do not know either of the terms *integer* or *multiple*. Adults know that the solution to $0 - 9$ is (-9). But eight-year-olds inhabit a mathematical world comprised solely of positive whole numbers. In this world, is $0 - 9$ impossible? Or is the answer zero?

Adding to the challenge for the elementary-grades teacher is not just answering reasonably but figuring out what is required for children to reason about mathematics in school (Ball & Bass, 2003; Lampert, 1990, 2001; O'Connor, 2002; Sfard, Nesher, Streefland, Cobb, & Mason 1998; Yackel & Cobb, 1996). Facing this challenge requires consideration of what mathematics is, offers, and requires. What constitutes mathematical reasoning? What even demands a reason? Facing the challenge also entails respect for the fact that the people involved are *children,* with mathematical knowledge that is both underdeveloped and under development. This reality raises the question, What version of mathematical reasoning is appropriate for children? Furthermore, this reasoning occurs *in school,* where norms exist for how students and teachers relate, and about what, and with what purposes and constraints. So in part this challenge demands consideration of mathematics, of students, and of school as a setting for learning mathematics. Additionally, and often less visibly, it also involves *teaching.*

---

4 This class period has been written about elsewhere (e.g., Ball, 1993).

5 Quotations of classroom talk from the episode under study are referenced throughout this chapter to the transcript found in Appendix 1 to this volume.

In our research, we seek to uncover and make visible what the *work of teaching* mathematics is; what makes the teaching, itself, *mathematical;* and what teachers do to *make* their work. To ground our inquiry, we analyze data from elementary school classroom teaching of mathematics. A primary resource for us is a large base of "records of practice" collected in Ball's third-grade public school classroom during the 1989–1990 school year.[6] A second source of material consists of analogous, albeit less extensive, materials from other teachers' classrooms and records from subsequent years of Ball's teaching.

The title of this chapter, "Making Mathematics Work in School," frames a central problem of our research. Using perspectives drawn from multiple disciplines—psychology, organizational theory, philosophy, linguistics, anthropology, sociology, and mathematics itself—we seek to uncover and analyze the work that teachers do. In particular, we do so with an eye to a fundamental question about how to treat the discipline of mathematics with integrity in the context of helping pupils learn. Following Dewey (1902/1990), Bruner (1960), and Schwab (1961/1974), we ask the meaning of being "intellectually honest" about the subject matter, of representing within the curriculum "fragments of the discipline," and of engaging pupils in fundamental elements of knowing and doing mathematics in school. As we study the work of teaching, we seek to understand its demands and the resources useful to its practice.

The words of our title evoke aspects of what teachers *do* in several senses. In one sense, the title refers to what teachers give students to *do* so that they will learn mathematics. Teachers define and *make the mathematical work* in which students engage: tasks, activities, and questions (Doyle, 1983). We investigate what that "work" is: What do teachers ask students to do that we would call the "mathematics work" of a lesson or a sequence of lessons? Certainly it is more than the assigned tasks (Arendt, 2000). Surely it includes the time, the materials, and what is done with them. It also includes the *talk* that carries and surrounds the tasks. That talk, and the work that constitutes it, is the focus of this chapter.

In a second sense, our title calls attention to the teacher's challenge of making it possible to *do mathematics in school.* We study what is required to maintain the integrity of the mathematics that students do in school. School mathematics is often remote from what counts as mathematics in the discipline. Its scope distorts the structures of the discipline, representing ideas as rules, problems as mere exercises, and solutions as fact-based answers (Goodlad, 1984; Lampert, 1990; Schwab, 1961/1978; Stodolsky, 1985). In this second interpretation, then, we explore what it might mean for mathematics to "work" in school. We consider what distinguishes a distortion of the discipline from a properly scaled, "mathematically honest" version, appropriate for 6-year-olds or for 10-year-olds.

---

6 These records were collected in Ball's third-grade mathematics class and Magdalene Lampert's fifth grade, with support from the National Science Foundation, for a project in which we set out to investigate the potential of using new technologies together with extensive records of practice to design new approaches to the pedagogy and curriculum of teacher education. For a discussion of this project and its results, see Lampert & Ball (1998).

In a third sense, the title points to teachers' putting mathematics to work so as to do the work of teaching. Mathematics can be thought of as a resource for teaching (Ball, 1993, 2000; Ball & Bass, 2000a; Lampert, 1990, 2001). As teachers work—to motivate students, to make sense of what students say, to provide a sense of closure for a lesson—mathematics offers ideas, language, and moves for carrying out that work. For example, asking a good mathematical question at the right moment can pique students' interest and draw them into the work. Knowing the "right" mathematics can help supply that question. In this interpretation of the title we ask, What does mathematics afford as a tool for carrying out the work of teaching? What mathematics is useful to know? And where in teaching is that mathematics likely to be helpful?

Finally, our title calls attention to teachers' core responsibility to ensure that what they do "works." No well-designed mathematical task or activity, nor amount of careful attention to the integrity of the mathematics, alone constitutes effective instruction. Rather, teaching works only if students develop mathematical proficiency, if they learn and grow, and if they become capable of doing and using mathematics and disposed to do so with confidence (Kilpatrick, Swafford, & Findell, 2001).

This chapter represents one part of our larger study of mathematics teaching. Like the companion chapters in this volume, we focus on this one day of mathematics class—in particular the short 6-minute segment from the beginning of the lesson. Across these 6 minutes, the teacher asks students to talk about their experiences of the meeting with the fourth graders. Although students might have chosen to talk about the dynamics, or what it was like to be with the older pupils, mathematics dominates their discussion. Students express ideas, offer mathematical explanations, and generate useful examples. They engage with one another, listen thoughtfully, and respond with interest. They agree and disagree. Observers of this episode often note students' serious engagement—with mathematics and with one another—and wonder what is required to produce it.

In this segment our attention was drawn to *talk:* to how the teacher's and students' work can be understood through a close examination of their talk—an examination that keeps its eye on both the discipline of mathematics and the practice of teaching. Although students do much of the talking during this segment, we sought in our analysis to uncover the teaching that structures students' mathematical talk. Our study of that talk led us to identify three aspects of the mathematical work in which students are engaged through that talk, and we used those three aspects to probe the work of the teacher. These three aspects are *naming and using names, making and interpreting claims,* and *evaluating mathematical assertions.* We show that they are, at once, both mathematical practices and teaching practices.

TALK IN MATHEMATICS CLASS

A 6-minute segment of classroom interaction can be studied in myriad ways. For example, one could study the written work done by individual students during a class segment, or pay close attention to the chalkboard work produced by the teacher and students for public record. Alternatively, one could obtain the curriculum and lesson plans for this particular day to analyze where the lesson conformed to, or deviated from, the intended plans. One could record which students participate, and document students' physical signs of engagement, making note of the socioeconomic status of each student. In this article, however, we turn our attention to the *talk* during this 6-minute segment while keeping our eye on the discipline of mathematics and the practice of teaching.

We made this analytic choice for several reasons. One simple reason is that talk pervades this particular slice of the day's lesson. Had we studied a different 6-minute segment from the same day's lesson, we might have chosen instead to analyze the individual written solutions children had produced when working at their desks. A second reason for our choice is that talk is thought to be a primary vehicle for student learning. Vygotsky's work (1978, 1986) and that of others (cf. Bakhtin, 1981, 1986; Wertsch, 1998) have emphasized the prominent role that speech plays in learning. A third reason grows directly from our own larger research agenda: Talk is a primary medium of teaching (Cazden, 1986). As one of teaching's major "technologies," talk is important to study. Particularly worth examining is teachers' own mathematical talk and how they elicit mathematical talk from their students. Significant analyses of classroom discourse (e.g., Au, 1980; Cazden, 1991) offer portraits of the discursive action and interaction of lessons. Still, less is known about the workspace in which teachers can deliberately cultivate and mediate classroom talk. A fourth reason for our interest in talk is our interest in equity. The facility with everyday talk that children bring to the classroom from home can be a platform from which to build mathematical talk. Perhaps more than other resources needed for mathematical work, children bring knowledge and experience of talk to the classroom. Elsewhere (Ball, Bass, Hoover, Lewis, & Wall, 2003), we have shown that the teacher's skillful design and ordering of questions can engage *all* children in the mathematical work of the class. With careful attention to the discontinuities as well as the connections between everyday talk and mathematical talk, teachers can use children's facility with everyday talk as a springboard to mathematical talk. Thus, although beyond the scope of this chapter, our decision to focus on classroom talk in our research is driven in part by our conjecture that although language can be a barrier to access, talk—used prudently—can also make mathematical work accessible to all children.

Our work joins a growing corpus of research on talk in elementary mathematics classrooms (see, e.g., Cobb & Bauersfeld, 1995; Kieran, Forman, & Sfard, 2002; Lampert & Blunk, 1998; Sfard, et al., 1998). For example, Yackel and Cobb (1996) consider how "sociomathematical norms" shape opportunities to learn mathematics for individual learners. They contend that "individuals' reasoning and sense-making processes cannot be separated from their participation in the interactive constitution of taken-as-shared mathematical meanings" (p. 460). Thus, their focus is on the extent to which conversation enables learning for the individual student. O'Connor (2002) begins instead with the mathematical content that one teacher is likely to encounter, and from that base considers what might be needed for the teacher to plan, carry out, and review a "position-driven discussion" that will help students learn the target content. For O'Connor, the primary focus is the work of the teacher in orchestrating conversation among students that will help them learn specific mathematical content.

Also concerned with talk, Nesher takes a different view of its role (in Sfard et al., 1998). She distinguishes among mathematical competencies engendered by conversation in mathematics class: are those conversations *about* mathematics, or are they *talking mathematics?* For Nesher, *talking mathematically* means describing "the world's situations with the formal models of mathematics" (p. 43). Similarly, Sfard (2001) argues that mathematical thinking *is* communication, so mathematics *is* mathematical talk. Nesher, Sfard, and others distinguish natural language about mathematical ideas from actual mathematical talk that employs the symbols and language specific to the discipline. Their work calls attention to the sense in which talking mathematics is central to *knowing,* not just learning, mathematics.

Common to many of these analyses of classroom talk is attention to the social dimension of conversations in mathematics classrooms. Weingrad (1998), for example, uses politeness theory (Brown and Levinson, 1987) to analyze a fifth-grade mathematics lesson. She argues that students are engaged in a kind of social positioning and maneuvering that affects the academic work in which they are thought to be engaged. Many researchers see classroom conversations about mathematics as socially risky, in which children can, in Goffman's (1955) terms, "lose face." From this perspective, a significant problem is the dynamics of the social relations that underlie classroom transactions (Lampert, Rittenhouse, & Crumbaugh, 1996).

Our investigations have led us to appreciate the sense in which social and intellectual purposes can be productively intertwined in classroom talk. Drawing on Lampert's work (1990), we argue that the terms of participation can be reappropriated to "produce lessons in which public school students would exhibit— in the classroom—the qualities of mind and morality that Lakatos and Pólya associate with doing mathematics" (p. 33). Because classrooms are group settings, we see also the extent to which the collective nature of mathematical practice depends on and uses talk as a primary medium of mathematical work. Through talk, mathematical ideas are aired and examined, and then ratified, revised, or discarded. In sum, talk provides the connective resources for making mathematics work in the collective settings of school classrooms.

ANALYTIC FRAMING

To probe how classroom mathematical talk shapes the work of teaching, we began by examining the students' and teacher's talk across this 6-minute segment. Three questions framed our initial analysis:

1. What features stand out about the teacher's and students' talk? What do they talk about, with what tools, and to whom, and for what purposes?
2. What are the social and mathematical features of their talk?
3. What is the role of the teacher in prompting, supporting, and teaching students' mathematical talk?

We viewed and reviewed the video segment, and coded the transcript. We observed, in slightly over 6 minutes of videotape, five distinct chunks of student talk. Each chunk is punctuated by the teacher asking the students for comments about the meeting of the day before, and each request by the teacher is followed by a student-initiated comment that precipitates an exchange among two or more classmates (see Figure 1.2).

Turn number	Length of turn (in minutes)	Brief description
1	0:51	Teacher asks for comments about the meeting.
2–8	0:47	Sheena comments on the meeting.
9	0:02	Teacher returns to ask for comments about meeting.
10–20a	0:51	Sheena and Sean exchange ideas about whether zero is even.
20b	0:17	Teacher asks for comments about the meeting.
21–23	0:23	Mei comments on how the meeting changed her thinking.
24	0:03	Teacher asks for comments about the meeting.
25–58	2:26	Nathan describes how and why he changed his mind about zero's being even, and the class reacts.
59	0:08	Teacher asks for comments about the meeting.
60–66	0:52	Sean says that he thinks six is neither even nor odd.

Figure 1.2. Summaries of chunks of student talk in 6-minute segment from mathematics lesson.

Displaying the segment in this way, and examining each chunk of talk, led us to some observations. The teacher's purpose at the beginning of this class was to give students a brief opportunity to reflect on the meeting. We observed that students seemed to understand the task, and they responded to it, for the most part, by raising important mathematical points that had attracted their attention and that they were still pondering. Central among those points were the nature of the number zero, and, in particular, its status as even or odd. We noted that students who spoke were engaged with their classmates, listening to their comments and questions, and responding. We saw that the teacher spoke relatively less than the students, and yet she seemed to be substantially shaping their talk. The students' talk did not seem "natural"—that is, it did not seem like talk probable in a typical third-grade classroom. We wondered what lay beneath our observations. If such talk was indeed not automatic, then we suspected that it was being deliberately cultivated. We sought to uncover evidence of the teaching and learning that might be shaping the students' mathematical talk—and, hence, work.

Our further analysis led us to posit three essential elements that undergird the nature of the mathematical talk in the segment. One is the available lexicon for mathematical talk: what words and phrases are being used, and to name and label what things. We refer to this element as "naming." A second is the students' evident orientation to making and understanding mathematical assertions. Third are methods for evaluating such assertions—that is, for judging assertions to be true or untrue and for revising or discarding them. We turn next to examine each of those elements, exposing its role in the segment. Following this discussion, we consider the role played by the teacher in these three arenas and its impact on the cultivation of useful mathematical talk.

NAMING AND USING NAMES IN CLASSROOM TALK

Names shape the discursive space of classroom talk. Names specify what is being talked about, what ideas or processes are important, and who is to do what. When Sean introduces his observation about six to the rest of the class, he needs a way to name what he has noticed, and so he says that six can be "even and odd." Although his observation is not, strictly speaking, the conventional use of the terms *even* and *odd*, we can see in his talk the important need to name ideas. And, later, the coining of the term "Sean numbers" facilitates the students' consideration of properties of even numbers, odd numbers, and Sean numbers.

Observers of this third-grade class often note the vocabulary employed by both the teacher and students. They are surprised, for some common terms in use seem far from ordinary for young pupils. The students regularly inquire whether two *representations* are the *same,* make *conjectures* and *arguments,* request *proof, comment* and *disagree, confer* with classmates, and *revise* their thinking. Why do these students, over half of whom are learning to speak English as a second language, use such terms? What role does this lexicon play in their mathematics work?

Consider the opening of the segment in which Ball frames the task for the pupils (turn 1G):

Ball: I'd like you to be thinking back to yesterday and to the meeting that we had on even and odd numbers and zero. And I have a few questions. First—my first question is, I'd just like to hear some comments about what you thought about the meeting, what you noticed about the meeting, what you learned at the meeting, just what kinds of comments you have about yesterday's meeting? And could you listen to one another's comments so that we can um, benefit from what other people say? See what you—what you think about other people's comments?

This single turn is replete with terms that signal the objects of attention, the tools to be used, and the ways to work. One lexical set—*even and odd numbers and zero*—is explicitly mathematical. These three terms are used to identify the mathematical ideas under discussion. Additionally important for the pupils' focus is a *meeting* at which these mathematical objects were discussed. The term "meeting" identifies a forum in which many pupils came together to discuss even and odd numbers and zero. Other words specify what students are to do: Their assignment is to make *comments* and to *listen* to others' comments. "Comments" labels a kind of discursive turn in which observations or reflections may be shared. Making a "comment" is not to give an answer. Rather, it is to contribute a view, a perspective, or a reaction, for others' consideration—in this example, what was "noticed" or perhaps "learned." Hence, asking pupils for "comments" is to point to a particular sort of work. Similarly, they are to "listen" to others' comments. Wall's (2003) analyses of "listening" in these data show that student listening is an explicitly developed practice with perceptual, behavioral, anticipatory, and conceptual dimensions. He traces, from the first classes, how *listening* to one another's ideas was guided and encouraged. Wall's analysis shows that although listening involved courtesy, it was an intellectual—not merely social—practice.

A final interesting term in this single teacher turn is the word *benefit* (turn 1F):

Ball: Could you listen to one another's comments, so that we can um, *benefit* from
what other people say?"

This term is not likely familiar to eight-year-olds, particularly among whom such a high proportion are just learning English. It sounds "grown-up" and important and, in context, suggests important reasons for listening to others' ideas. Whereas the others seem to be well-defined terms, *benefit* seems less so. Instead, the teacher's use of this unfamiliar word may suggest something ineffable—a sense of respect and seriousness about the work in which the students are engaged.

Figure 1.3 displays the terms used in this single teacher turn, classifying them by the aspect of mathematical work that they name.

	Mathematical content	Mathematical practice	Learning activity	Attitude, stance
Even numbers	▓			
Odd numbers	▓			
Zero	▓			
Meeting			▓	
Comments			▓	
Listen			▓	▓
Benefit				▓

Figure 1.3. Aspects of terms used in Ball's beginning-of-class turn.[7]

---

7 By "mathematical content" we mean ways of doing that generate or work on mathematics. By "learning activity" we mean an instructional activity in this classroom that may or may not be mathematical in nature. By "attitude, stance" we mean a feeling or posture in the classroom.

Although all these terms were from one turn of the teacher's talk, similar patterns can be found in the students' talk. They use terms not likely to appear spontaneously in eight-year-olds' talk. And those terms shape the mathematical work they appear to be doing as reflected in what they say in class. Consider one of Mei's turns in this segment (turns 21–23):

Mei: Um, I h— I thought that zero was always going to be a even number, but from the meeting I sort of got mixed up because I heard other ideas I agree with and now I don't know which one I should agree with.

Ball: Um-hm. So what are you going to do about that?

Mei: Um, I'm going to listen more to the discussion and find out.

Figure 1.4 displays Mei's use of terms, which mirror her teacher's although she also uses terms not used by her teacher: *agree* and *discussion*.

	Mathematical content	Mathematical practice	Learning activity	Attitude, stance
Even numbers				
Odd numbers				
Zero				
Meeting				
Comments				
Listen				
Benefit				
Agree				
Discussion				

Figure 1.4. Aspects of Mei's use of terms in one turn sequence.

What is Mei doing, as reflected in her "comment"? First, she seems to understand what is called for in making a *comment* in this class. She reflects on a new confusion she has experienced as a result of what she heard in the meeting. Second, her comment also reveals what she was doing the day before in class: Her confusion grows out of the work she did during the *meeting*, in which she apparently *listened* to other' ideas about the parity of zero, and those ideas challenged her own prior notions. Third, her comment contains evidence that she knows that *agreeing* is a mathematical practice, something for which evidence is needed, and she is aware that she can gain evidence by *listening* to a *discussion*. For Mei, a turn labeled "comment" provides her a space for reflecting on and articulating what she did and thought about zero. And throughout her "comment," Mei refers to significant objects and activities as she talks about her mathematical work. The names—*listening, agreeing, meeting*—offer her labels for those objects and activities.

Whereas Mei's use of terms reflects her teacher's, Ball also helps Mei make her comment at mid-turn, asking in relation to her fresh confusion about zero, "So what are you going to do about that?" This addition extends observation and reflection to a natural consideration of implications for the mathematical work at hand. In doing so, the teacher scaffolds other students' work in making comments as well, helping them fill out what they have to say. One striking example of supporting students in giving comments is in the first turn of the segment, turns 2–7, when Sheena is the first to respond to the teacher's request for comments:

Sheena: I—I—I liked it because, well, I like talking to other classes and, and when you talk to other classes sometimes it helps.

She stops, but Ball probes, "In what way?" and Sheena continues, "It helps you to understand a little bit more." Ball probes again, "Was there an example of something yesterday that you understood a little bit more during the meeting?"

Sheena: Well, I didn't think that zero was—zero, um—even or odd until yesterday they said that it could be even because of the ones on each side is odd, so that couldn't be odd. So that helped me understand it.

The teacher underlines and labels Sheena's comment:

Ball: Hmm. So y— so you thought about something that came up in the meeting that you hadn't thought about before? Okay.

Sheena: (*nods*)

In this sequence, the teacher seems to be helping Sheena understand and perform the work of "making a comment." Ball asks her to fill in more detail by asking for an "example" and points her toward reflecting on what she might have learned. Her interaction with Sheena provides a scaffold for what counts as a "comment" and what sort of work might be done to produce one, out of the child's experience of the meeting. In this exchange, the teacher's use of terms (*example, meeting, understand*) helps to show the sort of work involved in "making a comment." [8]

In addition to the work of making *comments,* which is explicitly named and taught by the teacher, the students' talk reflects their active engagement in figuring out how to think about whether zero is even or odd. They listen to ideas, make assertions, and assess arguments to determine whether they *agree* or *disagree* with them. And as they do so, the teacher labels what they are doing with terms designed to provide linguistic structures to support their work.

One example of this active naming occurs when Sheena shares that she used to think that zero was neither even nor odd until the meeting, but when she listened to another student's explanation using the number line, she thought differently: "It helped me understand it." The teacher labels what Sheena did (turn 20E):

Ball: Sheena commented that it was good to have the two classes together because she heard an idea that she hadn't thought about and it made her think about and even *revise* her own idea when she was in the meeting yesterday.

---

8 This kind of teaching work, in which a teacher sensitively scaffolds students' participation in the doing, talking, and learning of the discipline, is typically more familiar to teachers when teaching reading and writing, but it applies as well in teaching mathematics.

Offering the term *revise* as a label for changing one's mind on the basis of new evidence or insight accords significance to this important sort of thinking. With a name to label it, the practice is made visible and is accorded value.

Lampert (2001) argues that names accord importance and that they make particular ideas, skills, and habits usable by labeling them. In her analysis of a year of fifth-grade teaching, she identifies three mathematical practices among those worth naming and teaching. She describes how she sought to teach these practices to her fifth graders, and how she also named them to give them special status in the public discourse:

- finding and articulating the "conditions" or assumptions in problem situations that must be taken into account in making a judgment about whether a solution strategy is appropriate;

- producing "conjectures" about elements of the problem situation, including the solution, which would then be subject to reasoned argument; and

- revising conjectures on the basis of mathematical evidence and the identification of conditions.

These activities were important to teach *deliberately*. They represent the essence of mathematical activity in a way that makes it doable by 10-year-olds (Lampert, 2001, p. 66).

Lampert's book offers close detail of the ways in which she worked to establish a classroom culture that would support mathematical reasoning. The mathematical practices shaped the ways in which she formulated the tasks she gave to students, and she also used them to name and draw attention to important moves students made as they began working. They structured the mathematical work in which she engaged her students.

In the next sections of this chapter, we examine more closely what the students and teacher do as they talk to make, evaluate, and decide on the truth of mathematical claims.

MAKING AND INTERPRETING CLAIMS

In mathematics, the making of claims is central. In the well-known episode with which we began this article, Sean made a claim about the number six. Provisional claims are, in essence, the raw material for building justified mathematical knowledge. Once proposed, claims are revised, refined, justified, or refuted. We see the third graders engaged in those activities with the guidance of their teacher. And justified claims, or theorems, are the primary products of mathematical work. Hence, mathematical talk in classrooms should involve the making and reworking of claims. This lens offers another perspective from which to examine the interactions in our 6-minute classroom segment.

In this segment, 34 mathematical claims are made, 30 by the pupils and 4 by the teacher. For example, Sheena asserts that, on the basis of what she heard at the meeting, zero is an even number. Sean claims that it cannot be so. Later,

Nathan claims that zero is "special." Another twenty-some claims are made about mathematics, roughly half by the teacher and half by the students. For example, the teacher says the fourth graders have been thinking about zero for a long time and still disagree, suggesting that disagreement is to be expected and that some mathematical issues are not easy matters and require thoughtful, long-term consideration. And Mei says she is going deal with her confusion by listening more, suggesting that mathematics is something one figures out by listening and thinking; it is an ongoing endeavor. Implicit in those claims is an image of what doing mathematics means.

Evident in the pupils' talk is recognition that both making mathematical claims and making claims *about* mathematics are central mathematical activities. When Sheena reports on her experience of the meeting, she recounts her reactions to claims that were made. She explains, "Well, I didn't think that zero was—zero, um—even or odd until yesterday they said that it could be even because of the ones on each side is odd, so that couldn't be odd. So that helped me understand it." Sheena recalls that before the meeting, she thought that zero was neither even nor odd, but that during the meeting, she heard an argument that caused her to change her mind. Revised, her new claim is that zero is even—or at least that it is not odd. What Sheena attended to at the meeting reveals her sense of the importance of formulating statements about mathematical objects, processes, and relationships.

Sean is similarly attentive to the role played by mathematical claims. He hears in Sheena's comment a claim that he doubts. To ascertain what she is saying, he suggests that he disagrees, and speaking directly to her, poses a question to probe what she has claimed: "What two things could make it?" Sheena is not sure she understands his question, and turning around to look at him, asks graciously, "Could you repeat what you said, please?" These two pupils' interactions in turns 10–19 are directed toward sifting through the various claims on the table. In putting out a mathematical claim, Sheena makes a mathematical move, creating space for further mathematical work. On this occasion she may not have meant to initiate a line of work, but her claim has that effect anyway. Sean picks it up and works with it. This process of making claims and responding to them lies at the heart of mathematical work.

In particular, across the 6-minute segment, we see multiple examples of children generating mathematical claims. Typically, in school mathematics, teachers and textbooks make claims and children try to remember them (Stodolsky, 1985). As Lampert. Rittenhouse, and Crumbaugh (1996) argue, for children to learn to formulate mathematical claims in school, the terms and structure of discourse will need to be reconfigured. And beyond that, a sense of what counts as mathematical work in school will need to be reconfigured as well. The students in this classroom have learned something about *what counts as a mathematical claim,* about *how to express mathematical claims so they are usable by others,* and about *how to evaluate and respond to mathematical claims.* Together, those activities do, indeed, reconfigure the mathematical work of mathematics class.

But what is involved in making and evaluating mathematical claims? Claims are part of everyday life. We make claims about the world around us, about how we feel, and about other people and the things they do. Claims represent our appraisals of the world. However, making *mathematical* claims in the classroom is often new territory for learners and teachers. Making mathematical claims requires developing a mathematical sensibility or mathematical eye. In many ways, mathematical claim-making launches mathematical work, but to launch productive mathematical work and productive learning, students and teachers need to understand what counts as a mathematical claim, what claims are worth making, and how, in a group setting, to express and respond to claims respectfully.

Of the 34 mathematical claims made in this 6-minute segment, roughly a third are claims about zero, a third are about making and combining even numbers, and a third are about the number six. For example, Figure 1.5 lists the claims proffered about zero.

Mathematical Claims About Zero
Zero is not even or odd.
Zero could be even.
Zero is not odd.
Zero has to be an even.
Zero is not an even number.
Zero is always going to be an even number.
Zero is not always going to be an even number.
Zero is even.
Zero is special.

Figure 1.5. Mathematical claims about zero.

This list highlights two central features of mathematical claims—the importance of precise language and the need to carefully clarify the meaning of claims. More than in other disciplines, mathematics demands precision. Precision is one of its hallmarks. Reading through the list of claims about zero, asking what is meant by each of those different claims, one can see why that precision is important. The differences are subtle, yet significant. "Is," "could be," "has to be," "is always going to be," "is not always going to be"—each of those expressions means different things. Each involves different lines of argument. When Sheena says, "It could be even because of the ones on each side is odd, so that couldn't be odd," what is she claiming? Is she saying zero is even, or that it is not odd, or does she see those statements as the same? And what is the role of the "could" and "couldn't" in her statement? To argue that zero *could* be even, one might focus on consistency (i.e., considering zero to be even fits with other agreements already made). To argue that zero *is* even, one might return to the definition for even numbers. Quantification and negation play prominent roles in formal mathematical logic as well as in less formal mathematical reasoning. And, as we see in these exchanges, the students are exploring this important mathematical terrain.

In part, the apparent breakdown in the exchange between Sheena and Sean stems from having reasoned unwittingly from different definitions, but the difference between "could be" and "has to be" also confounds their exchange. Their confusion may result from different interpretations of mathematical quantification, from competing conceptions of mathematical truth, or from their tentative sense of conviction. Whatever the reason, careful attention to the exact wording and meaning of the claim opens up an important line of mathematical work. As the pupils attempt to make and defend the claim, important tacit assumptions emerge. Some students think that even and odd are mutually exclusive categories, for instance, yet this idea is not shared by all.

With these claims about zero, we see that precision shapes and is shaped by reasoning. Precision matters, even in third grade. In turn, the need for precision heightens the need to clarify claims. And we see that as the students in this class make claims, they seem to engage in a process of clarifying those claims, whether on their own or with the support of the teacher. The major chunks of student talk in this 6-minute segment are, in large part, occasions of making and clarifying claims. Sean asks Sheena for clarification. Ball, along with other students, asks Nathan for clarification. And several members of the class ask Sean for clarification. Much of the mathematical work we see going on in the segment (a kind of work that the teacher scaffolds from the first day of class) is about clarifying, testing, and revising (further clarifying) the claims being made.

This dynamic is particularly visible in the next group of claims, in which Nathan says he thinks zero is "special" because "even numbers, like they they *make* even numbers" (see Figure 1.6). He claims that zero is "special," but he also claims something about the class of even numbers rather than about a specific number. The teacher articulates two possibilities for what he might be saying, "Were you saying that when you put even numbers together, you get another even number, or were you saying that all even numbers are made up of even numbers?" Betsy and other classmates continue the process of asking for clarification, and the variations they generate begin to clarify what Nathan is claiming.

In this progression of claims we see a process of claim-making and claim-clarifying, with proposed rewordings and revisions, with examples and counter-examples, with students' taking on Nathan's idea enough to put it in their own words and give it respectful audience. Claims like these, about classes of numbers, create their own challenges, and examples often play a large role. Nathan constantly turns to examples to indicate what he means, "like two plus two is four and four plus four is eight." Giving examples seems to be something these students have learned to do. Betsy counters Nathan with an example of her own, "You can't with six." These examples help clarify claims and help make them precise. By the end of the exchange, Betsy summarizes Nathan's thinking and Nathan agrees, "Yeah, I'm not going by every single number."

Mathematical Claims About Combining and Making Even Numbers
Even numbers make even numbers.
When you put even numbers together, you get another even number.
All even numbers are made up of even numbers.
Two even numbers, just that same.
All even numbers can be made up of two identical even numbers.
You can't with six.
Six is two odd numbers to make an even number.
You need three twos to make six.
Three is odd.
I'm just talking about, like two plus two is four, and four plus four is eight, and I just skipped six, so I just added the ones that, that add.
Six can be an odd number.
What you're doing is you're going by twos and then what two equals from then you go from—all the way up.
Yeah, I'm not going by every single number.

Figure 1.6. Mathematical claims about combining and making even numbers.

The last group of claims evident in this 6-minute segment, about the number six, returns to considering a specific number but then uses this example to define a class of numbers of which six is merely one. In part, these claims form a natural outgrowth of the work done across the first two sets of mathematical claims. The students' consideration of how six might be thought of as both even and odd engages them beyond this segment, eventually leading them to define "Sean numbers," those that consist of an odd number of groups of two (2, 6, 10, 14, ...).

In addition to making mathematical claims, the students and teacher in this segment make numerous claims *about* mathematics, as in Figure 1.7.

These claims reveal another layer of mathematical work. Making a mathematical claim is to state something about mathematical content. We note that the claims in Figure 1.6 are about odd and even numbers, the composition of numbers, and the methods for deriving conclusions about the status of different numbers. In contrast, the claims in Figure 1.7 are reflections on the process of working in mathematics, the characteristics of mathematics as a field, and the nature of experience in mathematics class. Making a claim about mathematics is to reflect on the practice of doing mathematical work, or the nature of the discipline of mathematics. Implicitly and explicitly, students and the teacher comment on mathematics and what it means to do it. When the teacher begins mathematics class by saying, "There's nothing to take notes about" and asks the students to think back to the meeting, she implicitly claims that a discussion at a meeting is legitimate mathematical activity and that reflecting on it and commenting on it are, as well. She communicates that usual

Explicit and Implicit Claims *About* Mathematics and the Doing of Mathematics
When you talk to others, it helps you understand a little more.
So you thought about something that came up in the meeting that you hadn't thought about before.
I disagree that zero has to be an even number.
Could you repeat what you said, please?
I disagree because if it was an even number, what two things can make it?
I could show you it.
That doesn't mean that it always is even.
You said it was [even].
This is still a problem—we didn't settle it—there's still a lot of disagreement.
The fourth graders have been thinking about this a long time and still disagree about it.
Sheena commented that it was good to have the two classes together because she heard an idea that she hadn't thought about and it made her think about and even revise her own idea when she was in the meeting yesterday.
What other comments do other people have about the meeting and what happened yesterday?
I thought that zero was always going to be a even number, but from the meeting I sort of got mixed up because I heard other ideas I agree with and now I don't know which one I should agree with.
I'm going to listen more to the discussion and find out.
First I said that, um, zero was even, but then I guess I revised so that zero, I think, is special.
You said something like that yesterday.
I disagree.
So what you're doing is you're going by twos, and then what equals from then you go from—all the way up.
Um, I don't have anything about the meeting yesterday, but I was just thinking about six.
That doesn't necessarily mean that six is odd.
We need in the group to have an idea that we're working with.

Figure 1.7. Claims *about* mathematics and the doing of mathematics.

mathematical activity in this class includes taking notes and that thinking is an important, ongoing part of doing and learning mathematics.

Several of the claims about mathematics in Figure 1.7 characterize mathematics as an exchange and a testing of ideas. For example, Sheena claims that talking to others "helps you understand a little more." In contrast, Mei points out that talking to others can also confuse: "I thought that zero was always going to be a even number, but from the meeting I sort of got mixed up." When the teacher asks her what she is going to do about becoming confused, she goes on to say, "I'm going to listen more to the discussion and find out." On the whole, students' comments imply that public testing of ideas is crucial to doing and learning mathematics.

What public accord means becomes more evident across the segment. At the outset, the teacher highlights the importance of hearing different comments from different people. She tells students to "listen to one another's comments" and to "think about other people's comments" so that they can "benefit from what other people say." Several times, she asks, "What other comments do other people have?" She expects different students to have different things to say and that those differences will contribute to the mathematical work at hand. The classroom segment gives evidence that students, too, value listening, thinking, and revising. Sheena, Mei, and Nathan explain how the meeting with the fourth graders was as an opportunity to hear and consider different ideas. In discussing the meeting, Nathan describes how he came to revise his ideas, "First I said that um, zero was even, but then I guess I revised so that zero, I think, is special."

On several occasions, students challenge one another's proposed claims. Doing so grows out of their listening and thinking. Such challenges often suggest the need for revision. And from their listening and thinking, students begin to "try on" other students' ideas and to express them in their own words. For instance, Betsy works hard to understand Nathan, raising challenges but also listening for what he is struggling to explain, "So what you're doing is you're going by twos…."

Across the students' claims about mathematics, we see the significant role mathematical reasoning plays in publicizing and testing of mathematical claims. For example, decisions are made on the basis of agreed-on definitions (as when Sean disagrees with Sheena's claim that zero is even, asking, "what two things could make it?" at turn 10); mathematical arguments are to be given in support of claims and to be interpreted by others (as when Sheena says she could "show you it," turn 14); and disagreement is to be expected and thoughtfully considered (as when the teacher explains that the class probably will not settle the issue and that "the fourth graders who have been thinking about this for a long time also disagree about it," turn 20).

These students are making claims about mathematics and how it is conducted. The point here is not so much whether their claims are correct but that they are engaged in discussing and commenting *about* mathematics. And they often do so by making claims about mathematics. Across the year, their ideas change and grow, but their claims about mathematics represent their maturing sense of what it means to do mathematics. Such claims reflect a significant component of the work this teacher does to engage her students mathematically.

The ability to make mathematical claims does not come into being magically. As we see in this segment, the teacher scaffolds the children's formulation of claims. The day before this one, the third graders and the fourth graders had had a meeting in which the topics of discussion were the number zero and even and odd numbers. The third-grade pupils observed older students making and evaluating mathematical claims, and this process constituted much of the mathematical "work" for that day. Part of the teacher's role in helping children formulate mathematical claims is to provide material that provokes thought. The meeting with the fourth graders offered such opportunities for thought.

In the segment we are studying, the teacher (a) provides mathematically engaging experiences for them to make claims about; (b) invites their claims about those experiences; (c) prompts them for sufficient elaboration of claims so that all can "dig into" them; (d) ensures that they hear one another's discussion of claims; and (e) solicits other students' reactions to the original claim. For example, in each of the four chunks of the segment, the teacher asks for comments. Each time, she queries the claims students make, even if briefly, first with Sheena, then with Mei, Nathan, and Sean. (And several times, students follow up with their own queries: first Sean probing Sheena, then several students probing Nathan, and finally, several other students probing Sean.) These clarifications allow students to "dig into" contributions and subsequently to size them up. When Sheena comments about the meeting, the teacher asks her to elaborate (turns 3–5): "In what way? ... Was there an example of something yesterday that you understood a little bit more during the meeting?" These probes draw out what Sheena has in mind and make her thinking understandable and usable by others.

The teacher also places importance on children's hearing one another, an important element in constructing the work of making, elaborating, and refining claims collectively. At several junctures, Ball stops to make sure students can hear and follow the conversation. For instance across turns 30–36, she questions Betsy about whether Betsy had "said something like that yesterday, too." When the fact that Betsy had not been listening becomes clear, the teacher does not reprimand her but asks matter-of-factly, "Were you not listening to this just now?" and she then summarizes Nathan's contribution so Betsy can respond.[9]

This talk *about* mathematics is a prominent part of the mathematical work going on in this class. And although it often lies beneath the surface, it is pervasive and frequently comes fully into view. Together, students' mathematical claims and their claims about mathematics provide the raw material for mathematical work.

Next we examine how students learn to evaluate the claims that they hear. Doing so, in turn, makes mathematical work productive.

Evaluating Mathematical Claims

In the previous section, we saw multiple instances in which the teacher not only solicited students' comments but also solicited students' reactions to a claim. For example, when she opens the class, Ball ends her comments with the request "And could you listen to one another's comments, so that we can um, benefit from

---

[9] For a thorough examination of the role of listening and hearing, and the teacher's role in promoting listening, see Wall (2003).

what other people say? See what you think about other people's comments?" After Sheena makes a comment, one that Ball treats as a claim, she asks for other comments from other children in turns 9–10:

Ball: Other people's comments? Sean?

Sean: Um, I—I—I just want to say something to Sheena, when sh— what she said about um that, that one, um—zero has to be an odd, an even number bec— I disagree because, um, because what what two things can you put together to make it?

Notice that Sean responds to the teacher's request for more comments by responding to Sheena's claim, taking up, in a sense, the teacher's original invitation to think about, and comment on, other people's comments and claims. Later in the segment, at turns 26–29, Ball also models this type of response in her own engagement with Nathan's claim:

Ball: Can I ask you a question about what you just said? And then I'll ask people for more comments about the meeting. Were you saying that when you put even numbers together, you get another even number—

Nathan: Yeah.

Ball: —or were you saying that all even numbers are made up of even numbers?

Nathan: Yes, they are.

Mathematical knowledge is built from initial—and increasingly improved—mathematical claims. In *Proofs and Refutations* (1976), Lakatos provides a heuristic for the construction of mathematical knowledge. He identifies a pattern: generating a mathematical claim, mustering support for the claim through the provision of reason, generating possible counterclaims, and ultimately revising the claim.[10] This pattern offers a lens with which to scrutinize the mathematical work in which the students and their teacher were engaged. Given the prominence of mathematical claims in the discourse, we ask, What becomes of these productions? What makes them productive?

A cursory examination reveals that the students' talk is saturated with the evaluation of mathematical claims. They evaluate claims by agreeing and disagreeing—by supplying confirmatory examples and offering challenging counterexamples, by asking questions and trying out particular claims. When Sheena says that she now thinks zero "could be" even, Sean, on his own, poses a question that presses the implication of her claim and seeks to show that it is false: "What two things can make it?" When Nathan announces that zero is "special" because "even numbers make even numbers," his classmates press on the claim by asking questions, disagreeing, and raising problems with the idea. Over the next several minutes, his

---

10 This rephrasing of Lakatos uses synonymous terms more prevalent in elementary mathematics education. Lakatos describes this pattern as (1) primitive conjecture; (2) proof ("a rough thought-experiment or argument, decomposing the primitive conjecture into subconjectures or lemmas"); (3) global counterconjectures; and (4) proof re-examined: "the 'guilty lemma' to which the global counterexample is a 'local' counterexample is spotted" (p. 127).

claim is challenged, and ultimately sharpened: He is going "up by twos"—not "by every single number"—and what he is claiming is true even for zero.

Notable across these and the other examples is how entirely comfortable the children seem to be about receiving public disagreement and how attentively they listen to it and respond. Despite the fact that they are engaged in conversations that involve substantial disagreement and agreement, students' tone and manner are strikingly dispassionate and matter-of-fact. In this segment, students do not shy away from disagreeing with classmates' assertions. In fact, the practice is so pervasive that it appears to be standard practice for mathematics work in this classroom. How is it that students comfortably sustain mathematical argumentation despite the fact that disagreement is usually an uncomfortable or disallowed social position in school (Lampert, Rittenhouse, & Crumbaugh, 1996)? What makes this atmosphere possible?

Consider the ways in which students are actively engaged in evaluating mathematical claims during this 6-minute videotape segment. In the long stretch of talk that follows, almost every child's turn represents agreement or disagreement with Nathan's claim that "even numbers make even numbers." Nathan's assertion arises as part of an ongoing discussion about the parity of zero. Claiming that "even numbers make even numbers," he marks zero as "special." Implicit in his argument is that, whereas $2 + 2 = 4$, and $4 + 4 = 8$, and so on, in each instance producing a new even number, $0 + 0 = 0$. Therefore, zero is "special." He implies that this "specialness" exempts it from classification as either even or odd.

Following his assertion, and before other students comment, Ball first seeks to clarify his claim. She asks, "Can I ask you a question about what you just said? . . . Were you saying that when you put even numbers together, you get another even number, or were you saying that all even numbers are made up of even numbers?" He nods: He says he is arguing that even numbers are made up of other even numbers. Ball motions to Betsy, reminding her that she, too, had made a similar claim the day before.

This phase of clarification of claims marks an important aspect of the passage from Nathan's private idea to a public claim. Before the claim can be properly evaluated, the claim itself is clarified. Ball restates Nathan's clarified claim and presents it to the class to evaluate.

In this following segment, the teacher continues to help clarify the claim as the students discuss it:

34. *Ball:*   Nathan said a minute ago that when you put even numbers together you get an even number.

35. *Betsy:*   Mm-hm.

36. *Ball:*   But he also said, I think, that all even numbers are made up of other even numbers.

37. *Mei:*   I disagree.

38. *Sheena:*   (*Says something to Mei.*)

39. *Ball:*   Two even numbers just the same.

40. *Nathan:*   Unh-uh.

41. *Ball:*   The same even number?

42. *Nathan:*   Yeah, like four.

43. *Ball:*   Like eight is four plus four? Are all the even numbers—can you do that with all the even numbers? That they'd be made up of two identical even numbers?

44. *Sean:*   Not—not—not—

45. *Betsy:*   (*Looking toward Nathan*) You can't. Like six. Six is two, two, ... Six you can't get two.

46. *Sean:*   Six is two odd numbers to make an even, to make an even number.

47. *Mei:*   Three and three —

48. *Betsy:*   (*Still looking toward Nathan*) You need three twos to make six. You can't put a four and a four or a . . .

49. *Sean:*   Three twos???

50. *Betsy:*   (*Looking toward Nathan*) Three's—Three is odd.

51. *Sean:*   Or, um—

52. *Nathan:*   I know that, but um, um I'm talking about like two plus two is four, and four plus four is eight and I just skipped the six so I just added the ones that, that add. Like the two plus two is four, and four is an even number, and I'm just talking about the things that, um, like—

53. *Sean:*   Six can be an odd number.

54. *Nathan:*   —what I just said—the um, like two is plus two is four and four plus four is eight and—

55. *Betsy:*   So what you're doing is you're going by twos and then what two equals from then you go from—all the way up.

56. *Nathan:*   Yeah, I'm not going by *every* single number. Like—

57. *Betsy:*   Okay.

58. *Nathan:*   —two, four, six, eight.

Although "I disagree" is uttered only once in this passage, 13 of the 25 turns in this passage imply agreement or disagreement with Nathan's claim. In turns 37, 45, 46, 47, 48, 49, 50, 51, and 53, students are disagreeing with Nathan's claim.

As students seek to evaluate claims, their moves to agree or disagree are based on three basic activities: (1) clarifying the claim or the reason underlying a claim, (2) providing examples to prove or disprove a claim, and (3) considering whether a claim would be true in *all* cases. The diagram in Figure 1.8 shows the structure of this work.

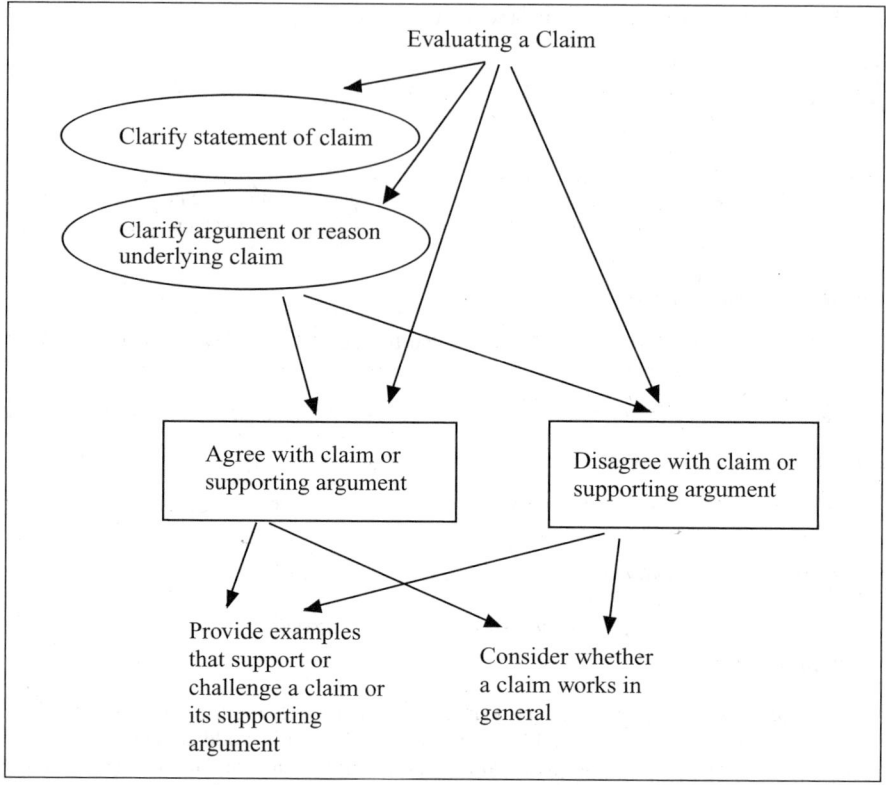

Figure 1.8. Structure of activities involved in evaluating claims.

Contrast this structure with the model of teacher decision-making offered in this volume by Schoenfeld (chapter 2), or the analyses of classroom talk offered by Horn in chapter 3 and by Posner in chapter 4. The structure offered here views teacher and student actions in the framework of a mathematical practice, the evaluation of a claim. In this model, the teacher's actions and the students' participation are framed by a mathematical practice that is social as well. It is social simply because evaluating a claim necessarily involves another person's idea and is conducted in the company of others. To compare this model with those in other chapters in this volume, we see that Schoenfeld's model is not specific to mathematics. Rather, his model of teachers' decision-making could ostensibly cross content domains. Horn's analysis concerns the learning of mathematics in the context of negotiating the social terrain; we suggest that social positioning drives Horn's argument. Posner's analysis foregrounds concerns for equity.

We return to our elaboration of Figure 1.8. In the short excerpt that follows, agreement or disagreement is bolstered by examples that prove or disprove the claim. The claim is considered at turns 45–50 by trying examples of numbers that either make the claim true or make the claim false.

Betsy: (*Looking toward Nathan*) You can't. Like, six. Six is two, two, ... Six you can't get two.

Sean: Six is two *odd* numbers to make an even, to make an even number.

Mei: Three and three—

Betsy: (*Still looking toward Nathan*) You need three twos to make six. You can't put a four and a four or a . . .

Sean: Three twos???

Betsy: (*Looking toward Nathan*) Three's—Three is odd.

A second path is to check for generalization. When Nathan says, "I'm not going by every single number," he takes a different path: a move to consider *all* cases. Beyond a single example that works to make the claim true, would the claim be true for *all* cases? This kind of move is central to mathematical reasoning, in which one checks the veracity of a claim for several cases and then seeks to ascertain that the claim is true for *all* cases.

A third pathway to agreeing or disagreeing with a claim is to clarify its underlying rationale. At several junctures, the teacher and the students clarify an idea previously offered in connection with the claim under consideration. Agreement or disagreement can be concluded through clarification of reason, as we see in this excerpt (turns 55–57):

Betsy: So what you're doing is you're going by twos and then what two equals from then you go from—all the way up.

Nathan: Yeah, I'm not going by every single number. Like—

Betsy: Okay.

Noteworthy is that the teacher's voice is nearly absent in the long passage—once she has launched this discussion by clarifying Nathan's two-part claim. After Ball clarifies Nathan's claim, Betsy immediately agrees ("Mm-hm."). The teacher adds that Nathan also thinks that all even numbers are composed of other even numbers, and Mei immediately says she disagrees. In the next more than twenty turns, the teacher's voice is not heard, and students offer examples and counterexamples to prove or disprove Nathan's two-part claim, and implicitly agree or disagree with his idea. Different compositions of the number six are offered as examples and counterexamples: three twos, three plus three. Students evaluate Nathan's claim on the basis of each example. Nathan is prompted by these examples and counterexamples to clarify his claim by bringing forth yet more examples, and mentioning that he does mean all cases ("I'm not going by every single number.")

How is it that students carry on this extensive evaluation of Nathan's claim without the teacher's interjection? This lesson is taken from a day in January, and students likely learned this form of mathematical argumentation over the course of the school year. But even within this short 6-minute segment of the lesson, we can see the teacher shaping the discourse toward this pattern of argumentation. Looking back to the beginning of the 6-minute videotape, recall that Ball had asked the class for reactions to the meeting with the fourth-grade class on a previous day. Note the pattern of Ball's queries, italicized in the text of turns 1E–7 that follows:

Ball: First—my first question is, I'd just like to hear some comments about what you thought about the meeting, what you noticed about the meeting, what you learned at the meeting, just *what kinds of comments you have about yesterday's meeting?* And could you listen to *one another's comments,* so that we can um, benefit from what other people say? See what y— what you think about other people's comments? Sheena, do you want to start?

Sheena: I—I—I liked it because, well, I like talking to other classes and, and when you talk to other classes sometimes it helps.

Ball: *In what way?*

Sheena: It helps you to understand a little bit more.

Ball: *Was there an example of something yesterday that you understood a little bit more during the meeting?*

Sheena: Well, I didn't think that zero was—zero, um—even or odd until yesterday they said that it could be even because of the ones on each side is odd, so that couldn't be odd. So that helped me understand it.

Ball: Hmm. So y— *So you thought about something that came up in the meeting that you hadn't thought about before?* Okay.

We note that the teacher begins with a question that anyone can answer, regardless of the student's mathematical knowledge. Responses to her request for reactions to the meeting can take almost any form, comprise any content. This opening question makes participation available to all.

Although the teacher is asking for general reactions to the meeting with the fourth graders, Sheena's reaction ("I liked the meeting") is treated as a kind of claim, and the teacher's questions that follow serve to clarify the reasons for her "claim," ("In what way?") and to provide examples ("Was there an example of something yesterday that you understood a little bit more during the meeting?"). Here the teacher establishes the pattern for evaluating claims: offer a claim, clarify the reasoning behind it, and offer examples in its support.

Sean then takes up Sheena's "claim" by disagreeing with her in turns 10–15, stating that her example does not work for all cases:

Sean: Um, I—I—I just want to say something to Sheena, when sh— what she said about um that, that one, um—zero has to be an odd, an even number bec— I disagree because, um, because what what two things can you put together to make it?

Sheena: Could you repeat what you said, please?

Ball: *(Speaks to Betsy and asks her to listen to this exchange.)*

Sean: Okay, um, I disagree with you because, um, if it was an even number, how— what two things could make it?

Sheena: Well, I could show you it. *(Moves toward the chalkboard and points to the number line above the chalkboard.)* Um, I forgot what his name was—but yesterday he said that this one (*points to the 1 on the number line*) and each—this one is odd and this one (*points to the –1 on the number line*) is odd, so this one has to be even.

Sean: But, that doesn't mean it always is even.

Consider closely the structure of the two students' arguments shown in Figure 1.9. Each student makes and evaluates a specific—and competing—claim about the parity of zero:

	Sean	Sheena
Claim	Zero is not even (but not that it is odd).	Zero is even.
Definition	An even number is a number that can be made up of two (equal) things.	Even and odd numbers alternate on the number line.
Prior knowledge		One and negative one are odd numbers (fourth graders said so).
Argument	What two *things* can make it (zero)?	Zero is situated between negative one and one.
	Implied: No two *things* can make it.	
Conclusion	Therefore zero is not even.	Therefore zero is even.

Figure 1.9. Mathematical structure of Sean's and Sheena's arguments.

Whereas Sean claims that zero is not even, Sheena claims that it is. Sean relies for his argument on a definition of even numbers that it can be "made up of two things." Sheena uses a different definition based on alternation on the number line. Although these two definitions are in fact equivalent, they are currently un-reconciled in the classroom discourse. Consequently, the two students' arguments reach opposing, contradictory conclusions.

With two different definitions, other mathematical work is needed to resolve the difference. Until the class has a shared definition for what an even number is, arguments such as this one cannot be reconciled. Ball does not attempt to engage in this significant piece of work on the spot. One does not always need to appease by resolving disagreements by fiat, without the proper tools to do so (Lampert et. al., 1996).

DISCUSSION OF MAKING MATHEMATICS WORK

In this 6-minute segment, the class engages in a mathematically sophisticated discussion, with students providing much of the mathematical substance. We analyzed what contributes to this outcome—what students do, what they have learned to do, and what the teacher does to teach them. We argue that a large part of what the teacher does is to reconfigure the talk that goes on and that this reconfiguring is directed at essential elements of mathematical reasoning—naming and using names, making and interpreting claims, and evaluating mathematical assertions.

Two symmetric acts combine to reconfigure talk in the mathematics classroom: On the one hand, the teacher's role is to ready mathematical content so that students can engage in it. On the other, the teacher readies students to be doers and learners of mathematics. We see in this 6-minute segment how the teacher weaves back and forth between readying the content for her students' use and readying her students to engage in it. On the surface mathematical content might seem to be "readied" by textbooks and curricula, but in fact teachers do a good deal to transform curricular documents into usable material for classroom work.

By way of example, consider the brief exchange, turns 26–28, between Nathan and the teacher:

Ball: Can I ask you a question about what you just said? And then I'll ask people for more comments about the meeting. Were you saying that when you put even numbers together, you get another even number—

Nathan: Yeah.

Ball: —or were you saying that all even numbers are made up of even numbers?

One can see in this excerpt how the teacher helps articulate Nathan's mathematical claim. To do so, the teacher has transformed a formal mathematical idea (that even numbers are composed of other even numbers) into a statement that third graders can manage ("when you put even numbers together, you get another even number"). The version she has readied for children lacks the rigor and precision of a formal mathematical statement. In mathematics, "putting numbers together" is not used to describe addition. One also does not "get" numbers. The teacher uses language to pose mathematical ideas in ways that are accessible to children.

At the same time, teachers also ready students to make profitable use of the mathematical material at hand. The videotape segment under discussion is replete with examples of the teacher doing so. In this same exchange, the teacher supports Nathan in practices that allow for mathematical work: she asks him to confirm his ideas with precision and clarity. She asks for his agreement, or for his adjudication. She models actively listening to his ideas, and publicly rendering them so that others can dig into them as well. In the segment we have studied for this chapter, *talk* is the agent through which the teacher attempts to mold these habits.

Talk in the mathematics classroom is a primary tool for doing this shaping of both the content and the students' work on it, whether it is rewording a problem midstream or recognizing the reasons for a student's halting statement. Thus, readying the mathematics for students' engagement, as well as readying the students to be doers and learners of mathematics, are largely a matter of designing and reconfiguring talk.

CONCLUSION

Mathematical reasoning is the foundation for the construction of mathematical knowledge. This chapter examines in detail the work entailed for teachers to engage students in this kind of fundamental mathematical work. Exhorting teachers to engage students in mathematical reasoning is inadequate as a support for their practice. Parsing the work of teaching makes instructional practice visible, and hence potentially learnable.

This article identifies three components of mathematical work: naming and using names, making claims, and evaluating claims. Evaluating claims, usually the focus of classroom analyses, is a complex and dynamic practice. Our analysis suggests that the "naming" of terms and definitions, and of ways of thinking and working together in the classroom, are foundational. We also suggest that such terms and labels are crucial for the formulation of claims. Teaching children to formulate a mathematical claim is not a straightforward endeavor: this article unpacks what is involved in such work. We then treat the evaluation of claims separately. Making analytic distinctions about the fostering of reasoning is essential to making teaching practice available for inspection and discussion.

The three practices—naming and using names, making and interpreting claims, and evaluating mathematical assertions—make visible what is involved in making mathematics work in the classroom. With these three practices in focus, teachers can reconfigure classroom talk for the purpose of engaging students in mathematical work. These practices offer ways of laying out the mathematical work to be done. They provide ways to focus on who talks, what gets talked about, and how it gets talked about. Our choice of them is strategic, serving both the discipline of mathematics and the social activity of teaching and learning. We close this chapter by describing three benefits of this resulting framework: epistemological benefits, relational benefits, and benefits for collective work.

Naming and defining, making claims, and agreeing or disagreeing are essential practices of the discipline of mathematics and foundational underpinnings to building understanding. As Ball and Bass (2001) argue, definitions and agreed-on language provide the base from which mathematical justification occurs. Reconfiguring these underlying features of talk from nondisciplinary discourse practices (such as typical patterns of teacher questions and student responses or justifications, for example, "My mother said …") to disciplinary ones (such as "I disagree because …" or "Nathan said a minute ago that …" or "What's our working definition?") provides a solid foundation for efforts to attend to teaching for understanding and the learning of concepts. It maintains the mathematical integrity of the talk and positions students to understand the mathematical basis for what they learn. The point here is not that students invent mathematics for themselves but that they come to see the mathematical basis for what is being taught and learned. For instance, what is the basis for knowing that six is not odd? As Mei concludes later in this lesson, if all numbers were both odd and even, we would have no reason to name or talk about odd numbers or even numbers. As students develop a sense for what is in a mathematical name and a mathematical justification, they become better positioned to understand the mathematics they learn.

Second, the framework we propose here provides resources for helping students see themselves as doers and learners of mathematics as well as for shaping relationships among students. In referring to Nathan's idea, Sean numbers, or "our working definition," students begin to identify themselves as people who do mathematics. Since names and claims can be connected, one with another—"Is that like what so-and-so said the other day?"—students' ideas get set in relationship with one another, and thus so do the students themselves. Likewise, agreeing and disagreeing gives students positions in the mathematical landscape and places students in relationship with one another. In these approaches, a teacher can take the words or ideas of students and make them central, putting students into the work in ways that are likely to help them see themselves as people who can do mathematics, perhaps against expectations and labels. These mathematical practices can be used to shape how students see themselves and how they are seen by others. Although mathematical argumentation is often seen as risky, in fact, arguing from a mathematical basis, together with revising, can be a resource for shaking up existing social dynamics (or setting up a world independent of those dynamics), and it provides mathematical criteria for respect in the mathematics classroom.

Finally, the framework furnishes resources for engaging students with one another in collective mathematical work. Names and definitions build a common language that can support collective work. They provide a base of common knowledge crucial for mathematical reasoning (Ball & Bass, 2000b), for instance, in establishing working definitions for even and odd numbers. Making claims is a kind of call to the group about work to be done. It sets the group agenda, focusing attention. By making a claim, one is in effect saying, "I take a stand, what do you think of this?" Making a claim invites people to engage in the mathematical work of evaluating mathematical assertions. This work provides a basis for knowing and learning, engages every student in the mathematics of other students, and engages them with other students.

Doing mathematical work in school depends on a reconfiguration of the talk, ubiquitous in classrooms, that is the medium of instruction. Sean did not serendipitously come up with his infamous claim about the number six; his classmates, in turn, did not engage seriously with him naturally. For students to learn to engage with mathematics and with one another, and for classrooms to nurture such engagement, requires, of teachers, work too often left invisible. By making the effort to unpack the work involved for teachers and their pupils, such practices can be learned and such mathematical work can become more the norm.

REFERENCES

Arendt, H. (2000). *The portable Hannah Arendt.* New York: Penguin Books.

Au, K. (1980). Participation structures in a reading lesson with Hawaiian children: Analysis of a culturally appropriate instructional event. *Anthropology and Education Quarterly, 11*(2), 91–115.

Bakhtin, M. M. (1981). *The dialogical imagination: Four essays.* Austin: University of Texas Press.

Bakhtin, M. M. (1986). *Speech genres and other late essays.* Austin: University of Texas Press.

Ball, D. L. (1993). With an eye on the mathematical horizon: Dilemmas of teaching elementary school mathematics. *Elementary School Journal, 93*(4), 373–397.

Ball, D. L. (2000). Bridging practices: Intertwining content and pedagogy in teaching and learning to teach. *Journal of Teacher Education, 51,* 241–247.

Ball, D. L., & Bass, H. (2000a). Interweaving content and pedagogy in teaching and learning to teach: Knowing and using mathematics. In J. Boaler (Ed.), *Multiple perspectives on the teaching and learning of mathematics* (pp. 83–104). Westport, CT: Ablex.

Ball, D. L., & Bass, H. (2000b). Making believe: The collective construction of public mathematical knowledge in the elementary classroom. In D. Phillips (Ed.), *Constructivism in education: Opinions and second opinions on controversial issues,* a Yearbook of the National Society for the Study of Education (pp. 193–224). Chicago: University of Chicago Press.

Ball, D. L., & Bass, H. (2003). Toward a practice-based theory of mathematical knowledge for teaching. In B. Davis & E. Simmt (Eds.), *Proceedings of the 2002 Annual Meeting of the Canadian Mathematics Education Study Group* (pp. 3–14). Edmonton, AB: CMESG/GCEDM.

Ball, D. L., Bass, H., Hoover, M., Lewis, J., & Wall, E. (2003). *Inattention to equity in teaching elementary school mathematics.* Paper presented at the Annual Meeting of the American Educational Research Association, Chicago.

Brown, P., & Levinson, S. C. (1987). *Politeness: Some universals in language usage.* New York: Cambridge University Press.

Bruner, J. (1960). *The process of education.* Cambridge, MA: Harvard University Press.

Cazden, C. (1986). Classroom discourse. In M. C. Wittrock (Ed.), *Handbook of research on teaching* (3rd ed.). New York: Simon & Schuster Macmillan.

Cazden, C. (1991). *Classroom discourse: The language of teaching and learning* (2nd ed.). Portsmouth, NH: Heinemann.

Cobb, P., & Bauersfeld, H. (Eds.) (1995). *The emergence of mathematical meaning: Interaction in classroom cultures.* Studies in Mathematical Thinking and Learning series. Hillsdale, NJ: Erlbaum.

Dewey, J. (1902/1990). *The school and society; and, The child and the curriculum/John Dewey.* Chicago: University of Chicago.

Doyle, W. (1983). Academic work. *Review of Educational Research, 53*(2), 159–199.

Goffman, E. (1955). On face-work: An analysis of ritual elements in social interaction. *Psychiatry, 18,* 213–231.

Goodlad, J. (1984). *A place called school: Prospects for the future.* New York: McGraw-Hill.

Kieran, C., Forman, E., & Sfard, A. (Eds.) (2002). *Learning discourse: Discursive approaches to research in mathematics education.* Dordrecht, Netherlands: Kluwer.

Kilpatrick, J., Swafford, J., & Findell, B. (Eds.) (2001). *Adding it up: Helping children learn mathematics.* Washington, DC: National Academy Press.

Lakatos, I. (1976). *Proofs and refutations: The logic of mathematical discovery.* New York: Cambridge University Press.

Lampert, M. (1990). When the problem is not the question and the solution is not the answer: Mathematical knowing and teaching. *American Educational Research Journal, 27,* 29–63.

Lampert, M. (2001). *Teaching problems and the problems of teaching.* New Haven, CT: Yale University Press.

Lampert, M., & Ball, D. L. (1998). *Teaching, multimedia, and mathematics: Investigations of real teaching.* New York: Teachers College Press.

Lampert, M., & Blunk, M. (1998). *Talking mathematics in school: Studies of teaching and learning.* Cambridge, MA: Cambridge University Press.

Lampert, M., Rittenhouse, P., & Crumbaugh, C. (1996). Agreeing to disagree: Developing sociable mathematical discourse. In D. R. Olson & N. Torrance (Eds.), *The handbook of education and human development: New models of learning, teaching, and schooling* (pp. 731–764). London: Blackwell.

O'Connor, M. C. (2002). Can any fraction be turned into a decimal? A case study of a mathematical group discussion. In C. Kieran, E. Forman, & A. Sfard (Eds.), *Learning discourse: Discursive approaches to research in mathematics education* (pp. 143–185). Dordrecht: Kluwer.

O'Connor, M. C., & Michaels, S. (1996). Shifting participant frameworks: Orchestrating thinking practices in group discussion. In D. Hicks (Ed.), *Discourse, learning, and schooling* (pp. 63–102). Cambridge: Cambridge University Press.

Schwab, J. (1961/1978). *Science, curriculum, and liberal education: Selected essays.* Chicago: University of Chicago Press.

Sfard, A. (2001). There is more to discourse than meets the ears: Looking at thinking as communicating to learn more about mathematical learning. *Educational Studies in Mathematics, 46*(1–3), 13–57.

Sfard, A., Nesher, P., Streefland, L., Cobb, P., & Mason, J. (1998). Learning mathematics through conversation: Is it as good as they say? *For the Learning of Mathematics, 18*(1), 41–51.

Stodolsky, S. S. (1985). Telling math: The origins of math aversion and anxiety. *Educational Psychologist, 20,* 125–133.

Vygotsky, L. S. (1978). *Mind in society: The development of higher psychological processes.* Cambridge, MA: Harvard University Press.

Vygotsky, L. S. (1986). *Thought and language.* Cambridge, MA: MIT Press.

Wall, E. S. (2003). *Making sense: Listening, remembering, and facilitating in the elementary mathematics classroom.* Unpublished doctoral dissertation, University of Michigan, Ann Arbor.

Weingrad, P. (1998). Teaching and learning politeness. In M. Lampert & M. L. Blunk (Eds.), *Talking mathematics in school: Studies of teaching and learning* (pp. 213–237). Cambridge: Cambridge University Press.

Wertsch, J. V. (1998). *Mind as action.* New York: Oxford University Press.

Yackel, E., & Cobb, P. (1996). Sociomathematical norms, argumentations, and autonomy in mathematics. *Journal for Research in Mathematics Education, 27*(4), 458–477.

Chapter 2

On Modeling Teachers' In-the-Moment Decision Making

Alan H. Schoenfeld
University of California at Berkeley

OVERVIEW

The preceding chapter, by Ball, Lewis, and Thames, documents the extraordinary complexity of teachers' work. Teachers have multiple goals, along multiple dimensions. They do work along the lines of elucidating mathematical content and processes. They do the work of creating and maintaining a classroom discourse community, in which the discourse reflects and respects a certain set of intellectual values. They do the work of understanding and relating to individual students, both with regard to their content-related understandings and with regard to their growth as human beings (see also Lampert, 2001). All this takes place in a dynamic environment in which classroom conversations can take unexpected twists and turns. As they work, teachers face a range of choices regarding possible actions, where pursuing one path may well mean forgoing others. One way to view such teaching, then, is as managing and resolving dilemmas (Ball, 1993, 1997; Ball & Wilson, 1996; Lampert, 1985). Seen this way, teaching can be regarded as an art form—grounded in principle, yet often spontaneous and opportunistic.

The issue addressed in this chapter is whether such in-the-moment decision making can be modeled, using the standard modeling tools of cognitive science. In this chapter introduction, I provide some background regarding the teacher-modeling enterprise and the reasons that the January 19, 1990, lesson represents an interesting and important test case for it.

For more than a decade, the University of California at Berkeley's Teacher Model Group (TMG) has worked toward developing a theory of teaching-in-context. Its goal is to provide a detailed theoretical account of how and why teachers make the choices they make while they are engaged in the act of teaching. Through 1996, most of TMG's work focused on analyzing mathematics and science teaching at the secondary and postsecondary levels. Detailed studies had been produced of Mark Nelson, a beginning teacher conducting a straightforward content-oriented lesson (Zimmerlin & Nelson, 1999); of Jim Minstrell, an experienced teacher conducting an innovative science lesson of his own design (Schoenfeld, Minstrell, & van Zee, 1999); and of Alan Schoenfeld teaching his college-level course on mathematical problem solving (Arcavi, Kessel, Meira, & Smith, 1998; Schoenfeld, 1998a). These studies, and the structure of a model of teaching, are described in Schoenfeld (1998b).

The claim made in Schoenfeld (1998b) is that the theory of teaching-in-context provides a general explanatory frame for understanding how and why teachers make the choices that they do in the midst of teaching, as a function of their knowledge, goals, and beliefs. In support of that claim, we should note that significant differences existed among the cases that TMG analyzed. The cases included beginning and experienced teachers, traditional and nonstandard lessons, and very different types of classroom interactions. In addition, huge differences in personal style were evident among the teachers studied. At the same time, the content, grade range, and classroom interactions spanned by those analyses (high school and college mathematics and physics) covered only a small part of the "teaching space"—that is, the range of teaching behaviors and contexts in general is much larger than the potential span of the examples that had been analyzed to date. Moreover, each of the lessons studied, even though wide-ranging with regard to content and classroom interactions, adhered closely to the teacher's agenda.

For many reasons, Ball's January 19, 1990, lesson is an ideal test case for the theory of teaching-in-context. Hers is a third grade class. All sorts of social, cognitive, and developmental issues that are not obviously present in high school or college students are central in this elementary school classroom. The teacher seems to play a much less direct role in shaping the course of events than do the teachers in the other lessons modeled by TMG. Indeed, the direction of the lesson seems to emerge as a function of what unfolds during the classroom conversations. Moreover, whereas each of the other lessons has a reasonably simple structure that corresponds to the teacher's intentions, this lesson (including the 6 minutes on which this analysis focuses) has so many twists and turns that no straightforward analysis appears likely to capture what happened. The challenge, then, is to determine whether the teacher's actions can be modeled. Success in the modeling enterprise would provide confirming evidence of the accuracy of the theory of teaching-in-context, and of the models derived from it; it would also help elucidate the knowledge and skills employed by the teacher during the complex act of teaching such a highly interactive, contingent lesson. Failure would help to specify the limits of the theory.

The structure of this chapter is as follows. I begin with a discussion of theories and models, and of the role that modeling a specific lesson segment can play in theory testing. I continue with a brief characterization of the theoretical perspective that provides the underpinnings of the work of the Teacher Model Group and a description of the representational form used by TMG to model the details of a lesson. Next I provide a brief narrative overview of the lesson segment in question. I then present an analysis of the lesson segment. The analysis unfolds in four phases. The first phase characterizes a simple iterative routine that can be used to debrief students on a topic. The second phase expands on this routine, describing a "flexible, interruptable" routine that can serve a teacher's debriefing purposes but can also be interrupted to pursue other high-priority goals. I claim that Ball's teaching during the first 6 minutes of the lesson corresponds, with great fidelity, to this second routine. The third phase provides the warrants for that claim. I analyze the transcript on a line-by-line basis, examining the decisions Ball made and the rationales underlying them. That analysis has three major goals. It seeks to indicate, first, that Ball's decisions and actions during the 6-minute episode do indeed correspond to the routine described in the second phase.[1] It further indicates that her choices, while made spontaneously, can be seen as principled and consistent in the light of her beliefs, goals, and knowledge. The fine-grained discussion of Ball's decisions establishes the basis for developing a model of her teaching during this lesson segment. In the fourth phase, I discuss the specifics of the model and the ways in which it represents aspects of Ball's teaching. This consideration, in turn, establishes the context for the concluding discussion. There I discuss the status and utility of models of the type discussed in this chapter.

ASPECTS OF A THEORY OF TEACHING–IN-CONTEXT AND MODELS THAT CORRESPOND TO IT

I begin with a brief introduction to the ideas underlying the theory of teaching-in-context developed by the Teacher Model Group. An extensive discussion of the theoretical enterprise and descriptions of several case studies can be found in Schoenfeld (1998b).

TMG's goal is to specify an analytic structure that enables one to characterize, in fine detail, the how and the why of an individual teacher's actions and decisions in the classroom. The structure is intended to be broad enough to characterize all teaching, from the most conservative instruction employing lectures or worksheets to the most radical attempts to foster highly interactive learning communities. The characterization of any particular teacher's actions (in a particular context) is presented as a model of his or her teaching (in that context).

On theories and models

The term *model* has multiple meanings in the education community. Among them one finds "to characterize the inner workings, in precise analytic terms" (as in *a model of working memory*), "exemplary" (as in *a model lesson*), and "to demonstrate" (as in *she will model the approach* in her sample lesson). TMG uses the term in the first sense. A model of a teacher conducting a particular lesson is a theory-based analytical characterization of the teacher's actions that explains

---

[1] An essential point to note is that the routine is one of many routines that Ball, a very accomplished teacher, has at her disposal. This chapter should not be misread as suggesting that this methodology is her only, or even primary, teaching routine; it is one she uses for a particular kind of situation, in which she wishes to uncover and explore the class's understandings of a particular topic. Evidence shows that this routine is far from idiosyncratic and is in fact employed by some other accomplished teachers as well; see Schoenfeld, 2002.

how and why the teacher made the choices he or she did while teaching. I use the term *theory* as defined in Webster's *New Universal Unabridged Dictionary:* "a formulation of apparent relationships or underlying principles of certain observed phenomena which has been verified to some degree."

Any modeling enterprise involves complex relationships among the theory that underlies the models, the models themselves, and the objects and relationships being modeled. The theory specifies a general class of relationships, delineating the components of models and the relationships among them. Working within this well-specified structure, researchers model phenomena of interest. The act of modeling serves as a dialectic mediating process between the theory and the phenomena being modeled. In one direction, a detailed theory-based model of a particular situation (in this instance, the teaching of a lesson or lesson segment) is intended to help develop a deeper understanding of what took place in that situation. In the other direction, the adequacy of the model-based description and explanation sheds light on the adequacy of the theory and serves as a mechanism for theory refinement. A failure of the model of an event to conform well to that event points to possible weaknesses in, or limitations of, the underlying theory.

To illustrate the relationship between theories and models, I discuss Sir Isaac Newton's formulation of a theory of gravitation. Newton posited that the gravitational attraction between two bodies is directly proportional to the product of their masses and inversely proportional to the square of the distance between them. In the abstract, this gravitational theory applies to any two objects, which may range in size from subatomic particles to supernovas. It is remarkably general.

How does one test such a theory? It cannot be tested all at once, but it can be tested on important subclasses of phenomena. For example, one might determine how well the theory does in characterizing the motion of planets within particular solar systems. Any specific solar system can be modeled using the laws of motion and the theory of gravitation. One can specify the positions, masses, velocities, and accelerations of the objects in the solar system at some particular time, past or present. Then one can run the model according to Newton's laws and predict the positions of those objects at various times after the start time. Doing so allows one to ascertain how well the model's predictions match the reality—that is, how good the theory is.

The modeling process fleshes out a theory in a number of ways. First and most important during theory development, it provides an empirical yardstick by which the accuracy of the theory can be assessed. If, for example, the use of gravitational theory results in the prediction of a lunar eclipse (or any other celestial event) and the eclipse does not take place, then one has good reason to be concerned about the theory! Accurate predictions, whether or not they are of practical import, serve as substantiating evidence that the objects and relationships specified in the theory are indeed correctly specified. Of course, substantiation using just one model is likely to be inadequate for a theory with wide-ranging applications; a range of models may be necessary to test the range and scope of the theory. In addition, models can have practical uses as well as theoretical ones. Once a theory is reasonably well established, practical applications tend to become increasingly important.

The same principles hold for theories and models of teaching. A general theoretical frame for examining teaching is as follows. Teaching entails knowledge-dependent decision making during a complex and highly interactive social activity. Any theory of teaching thus involves many of the features of a theory of human activity in complex social settings. Hence any theory of teaching must have much in common with general theories of cognition. The TMG's work is grounded in general cognitive theory, and its core components (representations) consist of a teachers' knowledge, goals, beliefs, and a decision-making mechanism. Its broad claim is that one can explain how and why a teacher does what he or she does in the classroom as a function of that teacher's knowledge, goals, beliefs, and decision making.

One way to explore the credibility of that claim is to try to construct models of different examples of classroom teaching. Thus, detailed models of specific instances of teaching play a very important role in examining the validity of the theory of teaching-in-context. Modeling specific cases tests whether the theory offers the tools to provide accurate descriptions and explanations of those cases. Modeling a range of cases (a beginning high school teacher working through a lesson covering straightforward procedural content, an experienced high school teacher guiding students through a very innovative lesson, a third-grade teacher cocreating her agenda with her students . . .) tests the scope of the theory. At the same time at least two kinds of practical implications are evident. The first is that the specific cases examined were chosen because of their intrinsic interest, and the models help explain just what the teachers were doing and why. In some instances the analyses point to pitfalls that might have been avoided; in others they point to skills or activity patterns that might not be apparent on casual (or even not-so-casual) observation. The second is that the modeling process can "unpack" some pedagogical routines that may have general utility. In this chapter I use a specific interactive routine to model Ball's teaching during the first 6 minutes of the January 19, 1990, lesson. The same routine turns out to be useful for modeling lessons taught by others, including Jim Minstrell and myself. Minstrell's lesson had some goals for student understanding and participation that were similar to Ball's but looked very different on the surface. Mine had slightly different goals, and looked different on the surface as well. This applicability suggests that the routine might be an effective way of achieving those and similar goals, and that other teachers might also learn to use the routine. For details see Schoenfeld, 2002.

The structure of the models we employ

The main theoretical idea is that what a teacher decides to do while engaged in teaching is a function of the teacher's goals (some of which are determined prior to the instruction and some of which emerge as the lesson unfolds), beliefs (which serve to re-prioritize goals as some goals are satisfied or new goals emerge), and knowledge (including various routines the teacher has for achieving various goals). The TMG's work provides a framework for describing and representing sequences of instruction. The full parsing of the lesson segment analyzed in this paper can be found in Appendix 2A. By way of introduction, Figure 2.1 provides a simplified version of Appendix 2A.

A transcript is given in the column at the far left of the representation. The lesson in its entirety is represented by one box in the column to the right of the transcript. The lesson is then parsed into "episodes," which are represented in Figure 2.1 by boxes numbered [1] and [2] in the next column to the right. The first major episode of the lesson represented in Figure 2.1, labeled [1], was the discussion of the previous day's meeting. It was followed by the class's extended discussion of Sean's conjecture that the number six can be both even and odd (the "evenness conjecture"). That discussion is labeled as episode [2]. Episode [3], not indicated in Figure 2.1, consisted of a classroom discussion in which the students reflected on the nature and purposes of their work. Indicators that each of these lesson segments is an episode at this level of parsing are (a) that each is coherent on its own terms (in this instance, each is "about" something) and (b) that the focus of each is different.

The next column in Figure 2.1 represents a decomposition of these episodes into smaller units. Episode [1] is parsed into five subepisodes, labeled [1.1] through [1.5]. These subepisodes represent an exchange between Ball and Sheena, a conversation between Sean and Sheena regarding the nature of zero, an exchange between Ball and Mei, an extended discussion of a statement made by Nathan, and a brief exchange between Ball and Sean. Since it is intended to be just illustrative, Figure 2.1 stops at this level of parsing. In general a full parsing continues down to the level of individual lines of dialogue uttered by the teacher and students.

To the right of the episode parsing, one finds representations of the teacher's attributed goals, beliefs, knowledge, and decision making. Goals are divided into three categories. Overarching goals (goals A, B, and C in this episode) are the consistent long-term goals that the teacher has for a class. They tend to manifest themselves frequently in instruction. They are often "in the background," increasing in activation when circumstances warrant. One aspect of Ball's teaching, for example, is that (Goal A) she wants to create a discourse community that, by virtue of respectful interactions, supports meaningful inquiry. A second (Goal B) is that Ball wants her students to engage in various kinds of specifically mathematical discourse, including making, clarifying, and evaluating mathematical claims. A third (Goal C) is that she has a strong desire to understand her students' thinking—she tends to question her students if she is unclear about their thinking and the circumstances seem appropriate. Major instructional goals (goals D and E here) may be oriented toward content or toward building a classroom community. Content goals tend to be more short-term, reflecting major aspects of the teacher's agenda for the day or unit. For example, Goal D is Ball's goal of "debriefing" the students regarding the previous day's meeting. It has extremely high activation during episodes [1.1], [1.3], and [1.5], and somewhat less activation during episodes [1.2] and [1.4]. Goals F through J are "local" goals, tied to

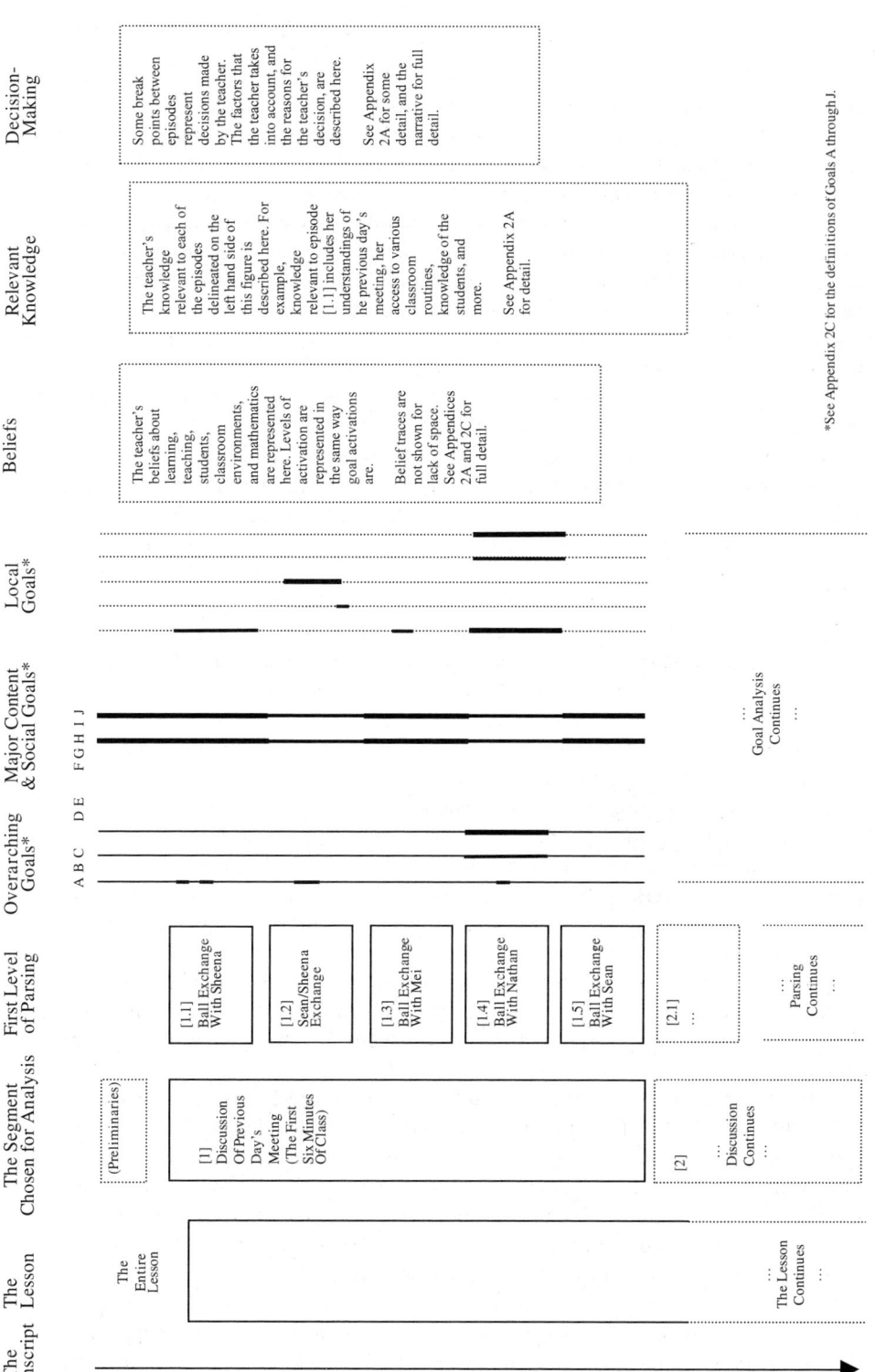

Figure 2.1. A simplified representation of the parsing of a body of instruction. The representation includes the transcript, the iterative decomposition of the instruction into episodes, goal and belief traces, and the teacher's relevant knowledge and decision making.

specific circumstances. Goal F, for example, is to have a student clarify, elaborate, or extend something that he or she has just said. Such a goal is obviously emergent (hence the dotted lines in Figure 2.1); the goal becomes active when a student says something that the teacher believes needs to be refined in some way.

Pointed to but not given in Figure 2.1 for reasons of space (but see Appendices 2A and 2C) is a delineation of the teacher's beliefs relevant to instruction and their intensity at various times. The full representation includes beliefs about learning, teaching, students, classroom environment, and mathematics. Beliefs are also represented as vertical bars of varying intensity. One "reads" their intensity the same way as for goals.

Following the description of beliefs is a delineation of the teacher's "relevant knowledge"—the many and diverse kinds of knowledge that a teacher can bring to bear during a classroom discussion. Such knowledge includes, of course, the classic categories: subject matter knowledge, general pedagogical knowledge, and pedagogical content knowledge. It also includes knowledge of classroom history—not only the previous day's meeting with fourth graders but also the topics the class has covered and how the discussions have gone. It also includes "other" history, for example, how last year's class dealt with some of these ideas. And it includes knowledge of the students—knowing, for example, that one student tends to call out responses and that another is shy. All this knowledge plays into the decisions that the teacher makes.

The teacher's decision making is represented in the final column of Figure 2.1. The reader should note that a teacher is always making decisions; some decisions are just more obvious than others. The clearest cases of decision making occur when some notable event occurs. For example, a student may raise a question. The teacher must decide whether to pursue it, and if so, how. Other less obvious decision points occur in the natural course of events, for example when a planned activity comes to an end. Should the teacher extend it, or move on to the next planned activity? Even if the lesson is unfolding largely as planned, myriad extemporaneous possibilities arise that he or she might consider. Whether and in what way the teacher decides to pursue any of those possibilities depends on circumstances and the teacher's knowledge, goals, and beliefs. Our attributions of the teacher's choices, and the rationales for them, are represented in this final column.

A representation such as the one outlined in Figure 2.1 can be used in two rather different ways. First, it can be used as a vehicle for description. One can parse a lesson, and then—using multiple sources of evidence, such as classroom observations, videotape analyses, the teacher's journals or other writings, and interviews to triangulate one's choices—delineate our attribution of the teacher's goals, beliefs, knowledge, and decision making. The picture that emerges is a rich, integrated portrayal of how and why the teacher did what he or she did.

Second and perhaps more controversial, the theoretical entities represented in Figure 2.1—goals, beliefs, knowledge, and decision-making mechanisms—can be used to produce models of teaching. To model a segment of instruction, one gathers various kinds of evidence regarding the teacher's goals, beliefs, and knowledge. One looks for consistency in the teacher's actions. This evidence tends to reveal both the teacher's knowledge structures (what routines does the teacher have for dealing with situations of type X?) and decision making (under which circumstances does the teacher tend to act in what ways?). Supporting data may consist of classroom observations, videotape analyses, the teacher's journals or other writings, and interviews with the teacher.

All these pieces can be put together as a model that simulates the teacher's decision making and actions. Let us call the model of the teacher MT. One starts with the teacher's "initial state": with what goals does the teacher enter the classroom, and what routines does he or she expect to implement to achieve those goals? This basis becomes the initial state of MT. That is, MT is assigned the teacher's knowledge, goals, beliefs, and decision-making mechanisms. Then a simulated lesson can begin, in which one works through the lesson by following the actions and decisions specified in the model. Working through the lesson consists of updating MT's knowledge, goals, and beliefs in response to events as they unfolded in the classroom, and having MT decide what to do at each point by having MT's decision-making mechanism act in response to its current knowledge, goals, and beliefs. What follows are two examples of how MT would make decisions.

Example 1. Suppose a lesson segment goes as planned. In that event, MT's goals for that lesson segment have been met and their activation levels are thus lowered. New goals corresponding to the next planned lesson segment become highest priority, and MT moves on to the next segment of the lesson just as the teacher would.

Example 2. Suppose that at some point in a lesson, a student makes an unexpected comment. How will MT deal with it? To make the situation specific, I shall briefly recap an episode discussed in Schoenfeld, Minstrell, and van Zee (1999).

The episode occurs toward the beginning of the term in a high school physics class. Eight students have measured the width of a table, obtaining values of 106.8, 107.0, 107.0, 107.5, 107.0, 107.0, 106.5, and 106.0 cm. The class has been considering how to determine the "best number" for the table's width. One student suggested the average, another, the mode; those alternatives have been discussed. Then a student makes this comment:

> This is a little complicated, but, I mean, it might work. If you see that 107 shows up four times, you give it a coefficient of 4, and then 107.5 only shows up one time, you give it a coefficient of 1, you add all those up and then you divide by the number of coefficients you have.

A priori, a wide range of responses is possible. One teacher might decide that this statement represents a deflection from his or her agenda, or that sorting through the answer would take too long. That teacher might quickly defer consideration of the issue ("Let's talk about it after class") and proceed with the lesson as planned. Another teacher might respond by showing the class that you have to divide by the sum of the coefficients, not the number of coefficients; and that once you realize that fact, the student's suggestion could be construed as a fancy way of computing the average. Yet another teacher might ask the class to think its way through the issue. Thus, teachers will differ in how they react; correspondingly, models of different teachers will differ in how they react.

In the situation analyzed in Schoenfeld, Minstrell, and van Zee (1999), the teacher being modeled had as a very high priority the creation of a classroom community of inquiry in which students are encouraged to make sense of situations. His pedagogical style was to respond to students' comments with questions that help to clarify issues but that put the responsibility for sorting through them on the students' shoulders.

In our analysis the model of this teacher, MT, was assigned the knowledge, goals, and beliefs we attributed to the teacher. MT had as a high priority understanding and honoring student inquiry; MT had a pedagogical style that made heavy use of questions as a way of engaging students and getting them to think through issues. Since the student's comment addressed the "best number" for the width of the table, MT judged it to be relevant. Moreover, MT had the knowledge that would enable it to understand the possible meanings and entailments of the student's suggestion (did the student mean to divide the sum of the numbers by 8, or divide the sum by 5?). The belief that relevant student inquiry should be encouraged was a very high priority for MT. Thus, even though exploring the issues raised by this student's comment involved a cost (in time), MT decided to explore them. But how? MT had to choose from among the various routines it knows. In this case, MT's goals, beliefs, and most frequently chosen teaching routine were all in sync: MT's preference is to ask questions that help clarify issues but put the responsibility for sorting through them on the students' shoulders. Thus MT asked the class what they thought of the student's idea (in effect, "What do you think? Does this make sense?") and had them work through it.

I shall now abstract this description to characterize the way that TMG's models make decisions in general. A new event (including the completion of planned activities) results in the activation of a new set of goals, beliefs, and knowledge, which, in turn, results in the activation of a set of plausible options (potential action plans) for what to do next. In the model, the choice among those action plans is made on the basis of a cost-benefit analysis. The model considers costs, benefits, and feasibility, as follows:

Costs:

If this action plan is implemented, what are likely costs in terms of—

(a) time?

(b) disruption to lesson continuity?

(c) pathways not taken, goals not achieved?

Benefits:

Which high-priority goals are likely to be achievable if this action plan is implemented? Might some serendipitous benefits be realized?

Feasibility:

What is the likelihood that the goals can be reached, given—

(a) the amount of time available,

(b) the material resources necessary and available,

(c) the students' readiness for this plan at this time, and

(d) the teacher's confidence in his or her ability to carry out this approach?

The decision is made on the basis of those considerations[2]. The lesson proceeds, and new events result in a continual updating of the current state.

Enough abstraction. The next section of chapter provides a narrative description of the first 6 minutes of the lesson in question. The following section models it.

THE LESSON SEGMENT IN QUESTION

This section presents a chronological narrative of the first 6 minutes of Deborah Ball's January 19, 1990, lesson, highlighting some of the decisions faced by Ball during the lesson segment and the ways in which she resolved them.

Ball's class was in the middle of a unit in which they had been exploring the properties of even and odd numbers. Ball had taught a similar unit the previous year. She observed this year that some similar issues had emerged in both classes, and also that some issues had gone unresolved for a long time in the previous year's class. (In fact, the previous year's class was still struggling with such issues as whether the number zero is even, odd, or "special.") For that reason she had arranged a meeting between the two classes. The meeting, which had taken place the previous day, provided a great deal of food for thought for Ball and for her students. Ball wanted to start off the January 19 lesson with a "debriefing"—what sense did the students make of the previous day's conversations? Her plan for the day was as follows:

> I had planned to spend a few minutes learning what the kids made of this op-
> portunity to talk about some rather large and perplexing mathematical questions
> with other kids. Among the things I wondered about is what they would make
> of hearing that the fourth graders also worried about and were interested in such
> similar issues, whether they would change any of their ideas from the discussion,
> and if so, on what basis, and what the overall intellectual and social experience
> was like for them. I wanted to return to the class's work on the conjectures about
> adding even and odd numbers because I was very excited at what looked to me
> to be one of the first signs that they would seek to prove a generalization through
> more than empirical data. They seemed on the verge of seeking to prove ana-
> lytically that an even number plus an even number would always equal an even
> number, and other related assertions. This is what I had planned to spend the day
> working on. (Ball, personal communication, April 4, 1999)

A transcript of the lesson segment is given in Appendix 1 of this volume. The comments in the discussion that follows are keyed to line numbers in the transcript.

---

[2] Some teaching decisions are toss-ups—in some circumstances a teacher might do A or might do B. In other situations, a teacher might be somewhat more inclined to choose A than B, choosing option A about 60% of the time. This tendency can be modeled by assigning probabilities to a range of possible outcomes. When the model is run multiple times, it will make different decisions, just as a teacher might.

Ball announces her intentions in Line 1: "I'd like you to be thinking back to yesterday and to the meeting that we had on even and odd numbers and zero.... I'd just like to hear some comments about what you thought about the meeting, what you noticed about the meeting, what you learned at the meeting." This comment sets the stage for her interaction with Sheena, who says (Line 2), "when you talk to other classes sometimes it helps." In Lines 3 through 7 Ball helps Sheena refine and articulate her claim. She abstracts Sheena's statement in Line 7, "So you thought about something that came up in the meeting that you hadn't thought about before? Okay."

In Line 10, Sean challenges Sheena's statement that zero "could be" even. He asks, "What two things can you put together to make it?"[3] The tension is palpable as Sean and Sheena interact in Lines 11–19.

The exchange between Sean and Sheena poses a series of dilemmas for the teacher. First, Ball's agenda is to have the students reflect on "what you noticed about the meeting, what you learned at the meeting." The conversation between Sean and Sheena focuses on the underlying mathematics (the nature of zero), not on the students' reflections about the meeting. Should Ball intervene to refocus the agenda, deferring a discussion of the content? Second, the tension in the exchange indicates that issues of ego and identity are at stake for both students. Should she intervene to avoid having Sheena feel bruised, or to defuse the tension? Third, various mathematical ideas are left up in the air as the exchange between the two students comes to a close. Sean and Sheena have been using two different approaches to explore the nature of zero—Sheena, the fact that even and odd numbers alternate; Sean, the working definition. Sean's definition actually does work: after all, two zeros do add to zero. Should one or both of those points be brought up, possibly deflecting Ball's agenda even further? Or should Ball allow things to settle, and move on?

Ball does not intervene. After the exchange ends, she takes advantage of what has been said to make some meta-level points regarding the fact that difficult issues take a long time to work through. She then asks the class for more comments about the meeting (Line 20). Her brief exchange with Mei (Lines 21–23) is entirely consistent with her agenda. In Line 24 she calls on Nathan.

Nathan responds, "Um, first I said that, um, zero was even, but then I guess I revised so that zero, I think, is special because, um, I — um, even numbers, like, they, they make even numbers; like two, um, two makes four, and four is an even number; and four makes eight; eight is an even number; and, um, like that. And, and go on like that, and, like, one plus one, and go on adding the same numbers with the same numbers. And so I, I think zero's special."

Ball might react to this statement in various ways. Nathan's statement, like Sheena's, could support a conversation consistent with Ball's announced agenda. Nathan began by saying that he "revised" his thinking—a term in this classroom's discourse community indicating that he had reconsidered his previous opinion.

---

[3] The class's working definition of even number was that a whole number is even if you can "make it" by putting together the same number twice. For example, $10 = 5 + 5$, so 10 is even.

His statement offers Ball a direct opportunity to pursue her agenda: she could ask Nathan what caused him to change his mind.

Nathan's comment could support a discussion of content, in the following way. One can rephrase one aspect of what Nathan said (but may or may not have meant), as follows: "Each even number, when doubled, makes a new even number. But zero, when doubled, does not—one gets zero again. Hence zero has a different property; it's special." This interesting argument might be pursued.

Nathan's comment also provides the fodder for a content-related conversation along different lines. Nathan observes that even numbers can be used to generate other even numbers, which can in turn generate other even numbers. Ball thinks (see Line 36) that Nathan has also stated the converse of the generative statement—that not only do even numbers make other even numbers but that every even number is made up of a pair of (the same) even numbers. Here, classroom history becomes relevant. Ball recalls (see Line 30) that Betsy had said something similar the previous day. Pursuing students' conjectures regarding combinations of even and odd numbers is part of Ball's agenda for that day. Pursuing it successfully depends on understanding where students stand with regard to them. Does Nathan or Betsy believe that every even number is double another even number? Do other students? Ball might wish to know the answer to those questions.

In sum, Ball faces a dilemma regarding possible reactions to Nathan's statement. She could pursue her announced agenda by responding to the part of his comment concerning the fact that he had "revised" his thinking. She could pursue the argument about the special nature of zero, thereby exploring an issue of mathematical interest to the class. She could ask Nathan to clarify whether he does indeed believe that every even number is twice an even number. Pursuing any one of these options will defer, perhaps indefinitely, the consideration of the other options.

Ball's choice, in Line 26, is to pursue the third, just discussed option: she asks Nathan whether he was saying that all even numbers are made up (of a pair of the same) even numbers. This consciously made choice puts her debriefing agenda on hold. In saying, "Can I ask you a question about what you just said? And then I'll ask people for more comments about the meeting," Ball signals that she is explicitly taking a detour from her intended path and that she expects to return to it once the detour is completed.

For purposes of subsequent discussion, the reader should note that Ball's decision in Line 26 could be questioned for a number of reasons. It might seem almost capricious. Why would Ball, who has carefully pursued her debriefing agenda, suddenly deviate from it? Having thus far steered the discussion away from specifics of content and toward issues of reflection, why would she now focus on content—indeed, on content that might be considered peripheral to the current discussion? If this decision is capricious, and if it is indicative of her style, it suggests a high level of arbitrariness in the decision making of a highly skilled teacher. If this decision is indeed capricious, the modeling enterprise might well be impossible. If this decision is not capricious, the challenge for the analytic enterprise is to show

how the decision is plausible and fits within a model of the teacher's classroom actions and decisions.

The exchange with Nathan, which lasts more than 2 minutes, concludes with an interaction between Nathan and Betsy that clarifies Nathan's stance regarding Ball's question. Nathan was not claiming that every even number is the double of some other even number, but rather that even numbers generate other even numbers by doubling.[4]

With this issue resolved, Ball returns to her agenda in Line 59, as she had promised 2 minutes earlier in Line 26. Her statement ("I'd really like to hear from as many people as possible what comments you had or reactions you had to being in that meeting yesterday") is direct and emphatic.

Sean's response in Line 60 to Ball's announcement represents yet another potential diversion from Ball's agenda. His comment appears to be related to the previous discussion of whether every even number (in this instance, six) is made up of a pair of the same even numbers. Ball and Sean have a brief exchange in Lines 61–66. During the exchange, Ball interprets Sean's comments as signifying that six is made up of two odd (rather than even) numbers, and she publicly confirms, with Nathan, that this interpretation is consistent with the stance he enunciated in Lines 56–58. From Ball's perspective, this confirmation responds adequately to the issue raised by Sean. In Line 67 she returns to her agenda, asking for more comments about the meeting.

In summary, the classroom segment on which this chapter focuses contains some very complex student–student and student–teacher interactions. Those exchanges represent teaching dilemmas, in the ways that Ball, Lampert, and their colleagues have used the term (Ball, 1993, 1997; Ball & Wilson, 1996; Lampert, 1985). I next turn to modeling the lesson segment.

A MODEL OF BALL'S DECISION MAKING DURING THE FIRST 6 MINUTES OF THE LESSON

The description that follows comes in sequential phases. My main claim in this section is that Ball's actions during the first 6 minutes of instruction can be modeled by the use of one complex decision-making routine. The routine specifies which decisions are to be made. The decisions are then made, through cost-benefit analysis, on the basis of Ball's attributed knowledge, goals, and beliefs. In what follows I start by characterizing a simplified version of the routine and "tailoring" it to model Ball by ascribing to it some of Ball's attributed knowledge, goals, and beliefs.[5] I next describe a more expanded version of the routine, which handles a greater range of classroom contingencies. I then provide a detailed analysis of the lesson segment, showing how Ball's actions during the first 6 minutes of the lesson correspond closely to five applications of the routine. I

---

4 One should read Nathan's statement in lines 56 and 58 as one continuous utterance. When one watches the tape, he is clearly saying, "I'm not going by every single [even] number like two, four, six, eight, . . ."

subsequently specify the model of Ball's teaching during this lesson segment by demonstrating the way in which the routine should be employed as part of the model. I then work through the model and show how it conforms to what actually took place during the lesson.

Before proceeding I should issue the standard caveats about the modeling enterprise. First, the point should be very clear that what follows is a model of Ball's teaching *during the first 6 minutes of the lesson.* I make no claim that this part of the model captures anything like the whole of Ball's teaching, or even the whole of her teaching during this lesson. The analysis focuses on one particular routine. Like any accomplished teacher, Ball has a wide range of pedagogical routines that might be employed to different ends at different times. Instead, I am claiming is that (a) this routine is a part of her repertoire; (b) she is likely to invoke it when she wants to find out what the students think about a particular topic; and (c) her actions in the 6-minute segment of tape are entirely consistent with its implementation. Similarly, the knowledge, goals, and beliefs specified in the model derived here are the knowledge, goals, and beliefs relevant to this segment of instruction. Some of them (e.g., the overarching goals A, B, and C listed in Appendix 2B and the beliefs listed in Appendix 2C) characterize aspects of Ball's teaching that I would claim are always present, but many of the other goals are "emergent" or tied to Ball's debriefing agenda; in other circumstances other goals would have highest priority. This discussion specifies only the knowledge relevant to this segment of instruction—a very small part of Ball's knowledge base.

Second, I want to emphasize the distinction between the form of decision making given in the model and Ball's actual decision making. In the model, decision-making is represented by a straightforward decision procedure. I make no claim that Ball, either consciously or unconsciously, employs this or a similar decision procedure as she teaches; indeed, such is unlikely. My claim, rather, is that working through the model produces decisions and actions quite similar to Ball's. Since the representations of Ball's goals, beliefs, and knowledge are as accurate as I can make them, my hope is that examining the model can produce insights into her teaching and the reasons for some of the classroom decisions she made.

At the core, a simple iterative routine

Here I describe the simple version of the routine that can be used to model Ball's teaching in this lesson segment. The core of the routine, in generic form, is represented as a flow chart in Figure 2.2.

---

5 The word *attributed* is important. It signifies that we as researchers assigned this knowledge and these goals and beliefs to the model on the basis of our analysis of Ball's journals, class notes, research papers, and videotapes of her teaching. As a form of research triangulation, TMG did not interview Ball about her knowledge, goals, and beliefs prior to developing the model. Part of our research strategy was to develop the model at least somewhat independently, and then to ask Ball, as a privileged informant, to comment on it. At the 1999 AERA session where we shared our analyses, she commented that the attributions appeared to be accurate. As a result of subsequent conversations, we have made some very minor changes in our analyses.

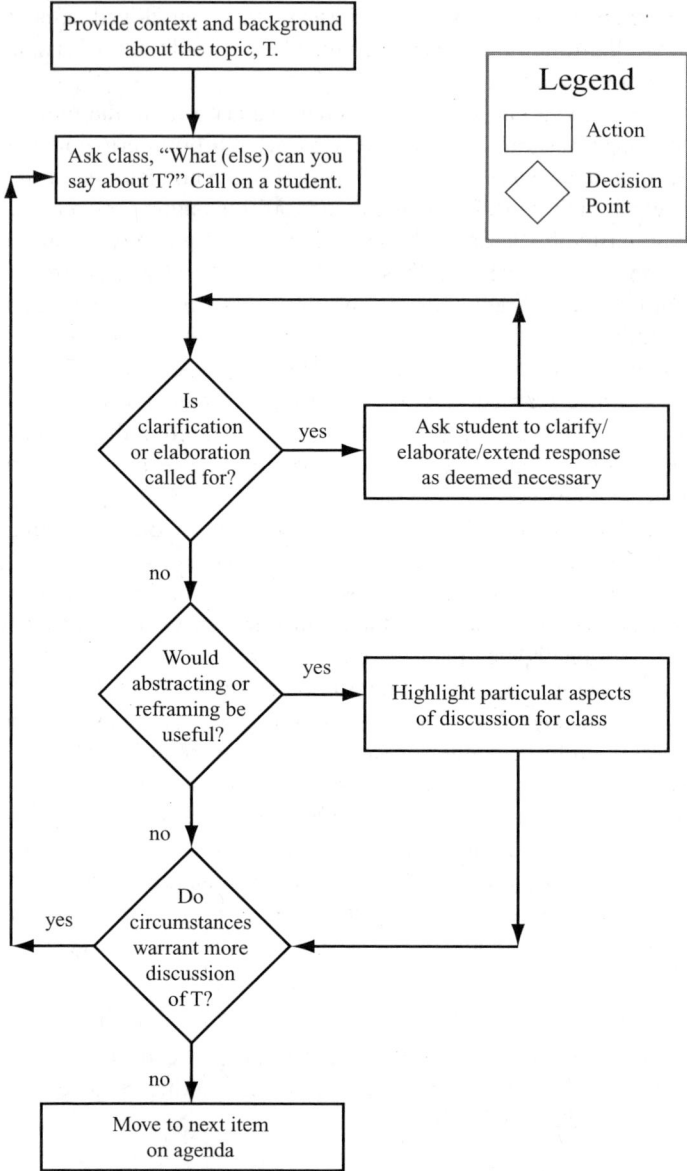

Figure 2.2. A simple routine for discussing a topic.

Note that even if Ball's teaching could be shown to match procedure in Figure 2.2 on a line-by-line basis, this description on its own is inadequate as a characterization of Ball's teaching. The routine as described in the figure is far too ill-specified to serve as a model because it fails to say on what grounds the decision at each of the decision points would actually be made. In its current form, this characterization of the teaching routine is generic—it could describe an extraordinarily wide range of behavior. At one end of the spectrum, for example, it could be used to characterize a conceptually oriented discussion in which the teacher, by means of subtle questioning ("What ideas do you think are relevant here? What do you think they are connected with? How does what you just said fit with what we were talking about earlier?"), helps students air and develop their understandings of a complex topic. At the other end of the spectrum, it could be used to characterize a purely skills-oriented lesson, in which the teacher's questions and comments ("What is the first step? Good. What is the second step? No, you do it this way....") are aimed at developing rote mastery of a procedure.

In other words, the major decision points in the procedure are all judgment calls. One's decision at each step depends on one's evaluation of the following: What calls for elaboration and what does not? Would summarizing, abstracting, or reframing be useful or appropriate at this point, and if so, what should the nature of one's comments be? Has this conversation run its useful course? Such judgments, of course, are grounded in one's knowledge of the students and of the mathematics, one's goals for the exchange, and one's beliefs about what is appropriate and possible. The model of a teacher must capture the teacher's judgments. Hence the generic description in Figure 2.2 becomes a model of a particular teacher only when it is tailored to the goals, beliefs, and knowledge of that teacher. The goals, beliefs, and knowledge must be specified clearly enough that one can work through the model and specify the decisions the model would make in response to classroom events as the lesson unfolds.

At this point, to begin to "personalize" the model of Ball's teaching, I shall provide a preliminary description of (our attributions of) some of her knowledge, goals, and beliefs as they are relevant to Figure 2.2. Much more detail regarding all these aspects is given in Appendixes 2B (a full parsing of the lesson), 2C (a list of relevant goals), and 2D (a list of relevant beliefs), and in the narrative that follows. Once Figure 2.2 has been fleshed out, I expand on it to place the iterative routine it represents in the larger context of Ball's teaching. The result, a flow chart given in Figure 2.3, will be used to analyze the first 6 minutes of the lesson in detail.

Goals

Ball begins the January 19, 1990 class with a very explicit goal for the start of the lesson. She announces this goal to the class: "My first question is, I'd just like to hear some comments about what you thought about the meeting, what you noticed about the meeting, what you learned at the meeting, just what kinds of comments you have about yesterday's meeting." As becomes clear in the interaction, she wants to have students reflect on the state of their understanding and on how they learn.

These goals sit within the context of Ball's overarching goals for instruction (see Appendix 2B). Ball wants her students to become mathematical sense makers who are inclined to think through complex issues and are capable of doing so. Her classroom community is the growth medium for such thoughts and dispositions, so she wants that community to be supportive of thoughtful explorations. This goal, in turn, has specific entailments, in terms of both community and mathematical norms. Norms [established by the teacher] are evident for interacting on an appropriately respectful basis (for example, students say, "I disagree," not "You're wrong"). Norms have been established for making assertions. Students' assertions are expected to be grounded in reason, and justifications for them are expected to be offered. (This norm is especially employed when one student's assertion contradicts a claim made by another student.) Mathematical conjectures are encouraged and explored. Ball has worked to establish these norms, and they are well established by this point in the school year; an ongoing goal is to maintain and strengthen them. In addition, as a teacher, Ball wants to understand her students' thought processes as well as she can.

Beliefs

Ball's beliefs about mathematics, students, teaching, learning, and classroom environment play a fundamental role in establishing her goals and shaping her choices of instructional actions. Many of these beliefs are listed in Appendix 2C. Of particular importance are Ball's beliefs that mathematics learning happens through—

- participating in mathematical discourse, which includes generating mathematical ideas, listening to others' mathematical ideas, and examining mathematical ideas, individually and collectively;

- participating in a community that values conjecture and reasoned argumentation; and

- articulating one's thoughts orally and in writing.

Such beliefs shape Ball's goals in clear ways. For example, the community norms that she tries to establish reflect her conceptions of what is important. They also shape her instructional decisions. In any given instructional context, Ball will act in ways that are consonant with her goals and beliefs. Indeed, some of her teaching routines have been developed as functional ways of achieving those commitments. Consider, for example, Ball's emphasis on having students learn to articulate their thoughts. One way to help students do so is to provide "scaffolding" for their explanations, gently nudging them toward expressing themselves with greater precision. An example of a scaffolding routine can be seen in the very first exchange of the lesson, with Sheena:

Ball: Sheena, do you want to start?

Sheena: I– I– I liked it [the meeting] because, well, I like talking to other classes and, and when you talk to other classes sometimes it helps.

Ball: In what way?

Sheena: It helps you to understand a little bit more.

Ball: Was there an example of something yesterday that you understood a little bit more during the meeting?

Ball's questions in this exchange provide a structure within which Sheena can clarify her own understandings and communicate them to the class. This general sequence of questions—first asking for clarification, then for an example—is a routine. Ball's choice of this routine is a reflection of her priorities, which are a reflection of her beliefs. The routine itself is part of Ball's knowledge base, discussed subsequently.

Knowledge

A full description of Ball's relevant knowledge would require volumes. Here I sketch some of her relevant knowledge and discuss the ways in which she accessed it.

Ball's knowledge of the content of third-grade mathematics is not at issue. The issues for Ball are (a) how her students might engage with the mathematics that they are studying as a community of mathematical sense makers and (b) what kinds of mathematical understandings she herself must have to facilitate such engagement. Ball's journals contain notes on the roles of definitions, conjectures, and mathematical argumentation. The very language of classroom discourse, including such statements as "I disagree because …" and "I revised my thinking," indicates the degree to which the class has been working on such issues. Those issues are very much at the forefront of Ball's thinking.

Ball's knowledge of where the class has been and where she would like it to go play major roles in determining her orientation to the lesson. Equally important is her knowledge of the students—who is shy, who is forthcoming, who knows what? The same holds for her knowledge of classroom history. Which issues have been problematic, which are "hanging in the air," which can be drawn on as threads for an interesting conversation?

Much of Ball's pedagogical knowledge is accessible in the form of routines, such as the scaffolding routine described previously. In too-simple terms, such routines are accessed when they are called for. For example, if you place high value on having students learn to articulate their ideas, and if the scaffolding routine is part of your repertoire, then Sheena's opening statement (Line 2) virtually begs for its application. To couch this scenario in modeling terms, Sheena's statement is a "triggering event" whose salient characteristic is that it is a vague statement made by a student. Given Ball's belief about the need for students to learn to articulate their thoughts with precision, she gives high priority to actions that will help Sheena clarify her statement. The "ask for a clarification/ask for an example" routine fits this need nicely and is selected. The general routine is then tailored to the particular circumstances. In part 1 of the routine, Sheena is nudged toward greater specificity when Ball responds to her comment "It helps" by asking her "in what way?" Sheena's response completes part 1, setting the stage for part 2, in which Ball asks for an example.

Other relevant knowledge is context-specific. For example, a student's comment such as "I don't understand" could, in the abstract, engender responses that range from "it's in your notes" to "I'll discuss this with you after class" to "let's work through this now." The choice of response will depend partly on context and constraints, for example, do we have time to explore this issue? The choice will depend on the teacher's sense of the class. Are other students likely to be confused? Would a discussion help? The decision will depend on the teacher's knowledge of the class history and goals. How does the student's comment tie into previous conversations? Can a response help lay the foundation for planned activities? It will also depend, fundamentally, on the teacher's knowledge of the student. Does the student often make such comments, or is this statement a rare expression of confusion? Is the student capable of dealing with a stern response, or is a "kid gloves" response appropriate? In short, the teacher's response depends on both general and very specific knowledge.

Despite the caveat in the previous sentence, an understanding of a teacher's general goals and beliefs does allow one to begin to specify the ways in which the routine described in Figure 2.2 could be used to model the teacher's actions and decisions. For example, one does not need much knowledge of Ball's teaching to assert that under typical conditions, she will not respond to a student's comment with evaluative statements of the type "that's right" or "that's wrong." Hence one can assign such statements extremely low response probabilities in the model. By contrast, Ball's beliefs and goals will, under typical conditions, lead her to respond to vague statements in particular ways. She is likely to try to help students articulate their thinking more precisely. Moreover, in certain conditions, she is likely to use a scaffolding routine of the type described previously. Hence, such responses can be built into the model. By doing so—by assigning each plausible response the probability that Ball would react in that way in that kind of circumstance—one can constrain the space of responses to the routine in Figure 2.2 to produce a rough approximation of the character of Ball's teaching.

A "flexible, interruptable" routine that includes debriefing at its core

The routine represented in Figure 2.2 is idealized. It represents classroom discussions as one might like them to happen, not necessarily as they do. In Figure 2.2 all the students' responses are implicitly aligned with the teacher's agenda, and the teacher works with those responses. In real classrooms, students do not always respond with comments that are aligned with the teacher's agenda. In such instances, the teacher has to decide how to respond. Sometimes a student's comment raises an issue that the teacher judges to be a candidate for classroom discussion. (Recall, for instance, the discussion of Example 2 in the section titled "The structure of the models we employ.") In that circumstance, the teacher needs to decide whether to follow up on the student's comment. If yes, the teacher must decide when and how to do so. Sometimes a student's comment points in directions that the teacher would prefer not to pursue. If so, the teacher moves to have the class return to the intended agenda.

Any model of a classroom routine must allow for such contingencies. Figure 2.3 represents an expansion of Figure 2.2 with such properties. The main difference between Figures 2.2 and 2.3 is a branch point in Figure 2.3 (the diamond labeled [D1]) that allows for a variety of options after a student has made a comment. This branch point represents the decision discussed in the previous paragraph. If the student's remark is in line with the teacher's agenda, then [D1] = "no" and the teacher proceeds to [D3]. From that point on, the teacher's interaction with the student is exactly as in Figure 2.2. If, however, the students' comment does raise other issues ([D1] = "yes"), the teacher needs to decide whether now is the time to pursue that issue. The choice is made at decision point [D2]. If [D2] = "yes," the class then works through the issue and (presuming a continuation of the main line of discussion is warranted) the teacher returns to the main topic. If [D2] = "no," the teacher moves to close off the discussion and return directly to the main agenda.

I claim that the routine outlined in Figure 2.3—supplemented with a great deal of detail regarding Ball's knowledge, goals, and beliefs—can be used to represent the first 6 minutes of her January 19, 1990, class with great fidelity. More specifically, I claim that Lines 1–67 of the transcript can be represented as five iterations of the routine outlined in Figure 2.3:

- providing context and background in Line 1,
- working through the "[D1] = no" pathway in Lines 2–8,
- working through the "[D1] = yes" pathway in Lines 9–20[C],
- working through the "[D1] = no" pathway in Lines 20[D]–23,
- working through the "[D1] = yes" pathway in Lines 24–58,
- working through the "[D1] = yes" pathway in Lines 59–66, and
- beginning yet another iteration of the routine in Line 67.

The analysis proceeds as follows. I first identify the main break points in the lesson segment. Doing so results in a first-level parsing of the lesson segment, which was previewed in the discussion of Figure 2.1. Each of the episodes in the first-level parsing is then analyzed in detail. Particular attention is given to decisions at branch points. The analysis indicates that Ball's choices, although made spontaneously during the flow of classroom events, can also be seen as principled and rational, in the sense that they represent choices that are highly consistent with her beliefs, and in the service of her high-priority goals.

A detailed analysis of the lesson segment

The analyses in this section and the next are keyed to the transcript, which is given in easy-to-read form in the Appendix 1 of this volume. The transcript also appears as the leftmost column of the parsing given in Appendix 2A. Appendix

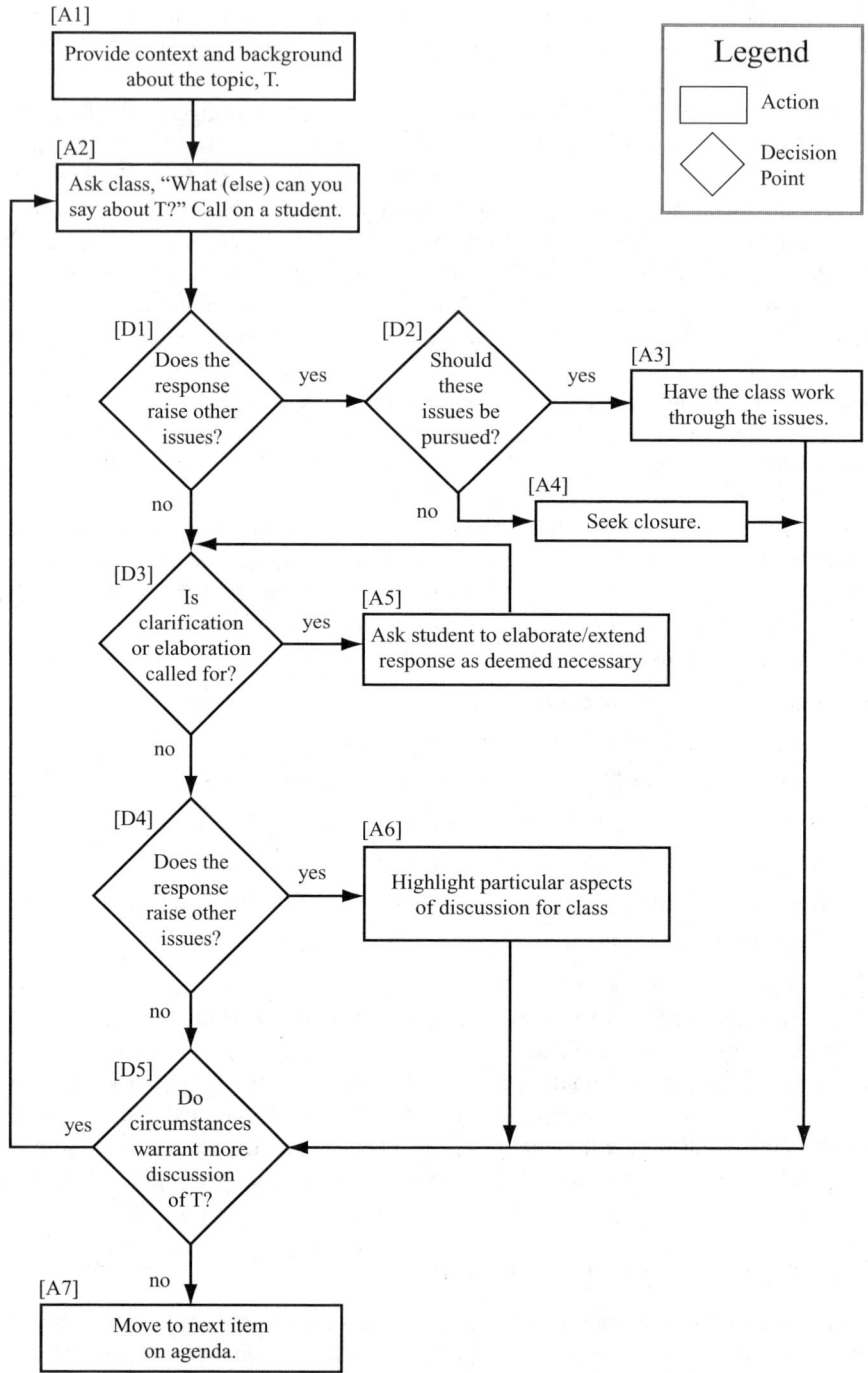

Figure 2.3. A flexible, interruptable routine for discussing a topic.

2A presents a relatively complete analytic representation of the first 6 minutes of the lesson. Unpacking the detail in Appendix 2A calls for a fair amount of cross-referencing with regard to goals, beliefs, and decision making. The goal traces in Appendix 2A indicate the intensity of Ball's overarching goals (A, B, and C), major content goals (D and E), and local goals (F through J) at any time. Descriptions of goals A through J are given in Appendix 2B. The belief traces in Appendix 2A indicate the intensity of Ball's beliefs about learning (L1 through L6), teaching (T1 through 4), students (S1 through S4), classroom environments (C1 through C5), and mathematics (M1 through M3). Descriptions of those beliefs are given in Appendix 2C. The rightmost column of Appendix 2A, "decision making," summarizes the decisions Ball makes and also gives the grounds for those decisions. Specific decisions are keyed to the decision points [D1] through [D5] in Figure 2.3.

As noted in the foregoing, my goal is to show that Lines 1–67 of the transcript represent five "tours" through Figure 2.3. Readers who want to follow the argument on a line-by line basis may find it useful to make copies of Figure 2.3 and Appendixes 2A, 2B and 2C, and lay them out in a way that facilitates cross-referencing.

I begin with a first-level parsing of the lesson segment. The general idea behind lesson parsing has been discussed: the subepisodes of any episode of instruction are chosen to be maximal subunits of that episode that cohere on phenomenological grounds. In more detail, the rules for parsing are the following:

a. When an episode is decomposed into subepisodes, each subepisode must cohere along some well-defined dimension, which may be the topic of discussion, the nature of the interactions (whole-class discussion versus seatwork versus individual presentations), or, at fine levels of grain size, turn-taking in conversation.

b. Although subepisodes of a given episode are not required to be of the same length (and may indeed differ substantially in length), they must be at the same grain size.

c. Distinguishing features must be evident between adjacent subepisodes.

d. Decomposition proceeds one level at a time, so that the subepisodes at the "next" level of parsing of any episode are the largest subepisodes of that episode that possess properties a, b, and c.

The first-order parsing of the lesson segment in question is straightforward. The first 6 minutes of the lesson split naturally into five episodes, as follows.

First episode. Ball begins (Line 1) by establishing the context for the discussion. Her exchange with Sheena in Lines 2–8, dealing with Sheena's reaction to the previous day's class, is focused and coherent. Ball's "Okay" in Line 8 effectively announces that that episode has come to a close.

Second episode. Ball's question "Other people's comments?" opens the episode. Sean's response begins a (rather tense) exchange between him and Sheena, which lasts through Line 19 of the transcript. Ball's comments [A] through [C] in Line 20 bring closure to that conversation.

Third episode. Ball once again invites comments in Line 20, and Mei responds. The brief exchange with Mei in Lines 21–23 is self-contained.

Fourth episode. In Line 24 Ball requests additional comments. Nathan responds in Line 25. The class discusses issues raised by his response all the way through Line 58.

Fifth episode. Ball reiterates her agenda in Line 59. Sean responds with a somewhat cryptic comment about six being made up both of two things (two threes) and of three things (three twos). Ball interacts with Sean and Nathan about this idea in Lines 60–66. Her invitation to the class in Line 67 ("Other people's comments?") indicates that the exchange with Sean and Nathan has come to a close.

Little more needs to be said about this decomposition. Each of the episodes coheres. Each is, in essence, the "working through" of a comment made by a student. In that way all the episodes are at the same grain size, even though some of them are much longer than others. Each episode is clearly distinguished from its predecessor. And, no coherent decomposition with fewer episodes appears to exist.

Context[6]

The story of January 19, 1990, begins before the beginning—of the lesson, that is. Ball enters the classroom that day with knowledge of what has preceded that day's lesson, and with a particular set of goals, beliefs, and action plans. Appendix 2B provides a list of some of Ball's relevant goals during the lesson.

Three types of goals are identified in Appendix 2B. Ball's *overarching goals* are always present at some level of activation. If particular classroom events occur, they may be "triggered" and take on very high priority. The *major content and learning goals for this lesson segment* establish Ball's orientation to the instruction in which she is about to engage. As noted in Section 2, Ball begins the class with a debriefing agenda. She wants to hear the students' reactions to the previous day's class session. Ball is specifically concerned with issues at the meta-level: she wants to orient the classroom conversation to issues of how the students understand their own mathematical learning and understanding. She will, however, let the discussion take its course as long as that course is consistent with other high-priority goals. The *local goals* identified in Appendix 2B are emergent. Such goals are triggered by specific student comments (e.g., a vague statement by a student triggers a "clarification" goal and results in the implementation of a clarification routine).

Ball's goals are fundamentally related to her beliefs about learning, teaching, and mathematics. Relevant beliefs are highlighted in Appendix 2C. They will be invoked as needed in the analysis that follows. So will relevant knowledge of students, curriculum, and classroom history.

---

6 For purposes of readability, the description in this section is given in the form of a narrative, including descriptive statements of the type "Ball has the following goals and/or beliefs." All such claims are assertions and should be seen as such. They should be interpreted as follows: "In this analysis, I attribute the following goals and/or beliefs to Ball." Such attributions are grounded in analyses of Ball's journals and annotated class logs as well as videotapes of her lessons.

Analysis of Episode 1

After some preliminaries, Ball begins the lesson (Line 1 of the transcript) by establishing the context for the day's discussion (item [A1] of Figure 2.3) and calling on Sheena (item [A2]). Sheena's response in Line 2, ". . . when you talk to other classes, sometimes it helps," is directly in line with Ball's agenda. The response does not raise other issues, so in terms of Figure 2.3, [D1] = "no." Sheena's response is vague, however; hence [D3] = "yes." The vague response activates Ball's clarification routine, represented by the subloop through [A5]. In Line 3, Ball provides a general prompt, "In what way?" Sheena's response in Line 4, "It helps you to understand a little bit more," is more specific, but more remains to be said. Thus [D3] = "yes" once again; in response, Ball asks (Line 5) for an example. The example Sheena gives in response (Line 6) is fully responsive to Ball's question ([D3] = "no"), bringing that subloop to a close. However, Sheena's response does provide the opportunity for Ball to make a meta-level point to the class ([D4] = "yes") in summary of what Sheena has said. Ball does so in Line 7 (Action [A6]). Then, since the debriefing activity has just gotten under way, Ball acts to continue it ([D5] = "yes").

Lines 1–8 of the transcript can thus be seen as one pass through the routine described in Figure 2.3. This pass is represented schematically in Figure 2.4. Detail regarding the ways that goals, beliefs, and actions fit together is given in Appendix 2A.

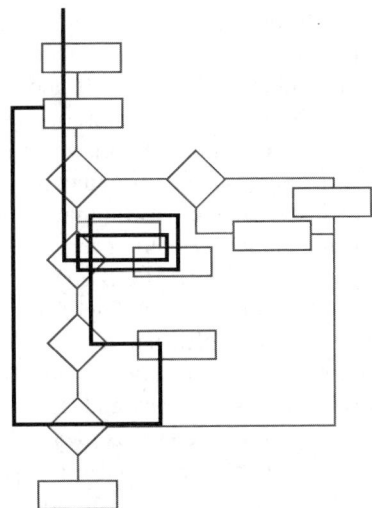

Figure 2.4. A schematic representation of the flow of lines 1–8 of the lesson transcript.

Analysis of Episode 2

Events take a different turn in Episode 2. In Line 9 Ball once again begins the iterative routine with a request for comments. She calls on Sean.

Sean does not discuss his reaction to the previous day's meeting (Ball's agenda), but instead addresses a content-related question directly to Sheena. His question in Line 9 implicitly uses a working definition for even numbers that the class has developed: an even number is always "made up" of two equal numbers. If zero is even, asks Sean, what two numbers make it up? Sheena responds substantively to Sean, although her response is at cross-purposes to his. In Line 14 she offers an argument that she had heard the previous day: even and odd numbers alternate, and zero lies between the two odd numbers -1 and $+1$, so it has to be even. Sean pushes: does it have to be even? Sheena backpedals in Lines 16 and 18, saying it could be even but doesn't have to be, at which point Sean notes (Line 19) that she had said earlier that it must be. The issue is left hanging, with some tension in the air. Sheena returns to her seat. In Lines 20A through 20C, Ball comments on the difficulty of such discussions and notes that even the fourth graders with whom they had met the previous day have not resolved the issue. In Lines 20D through 20G, Ball announces an explicit return to her debriefing agenda and calls on Mei.

The foregoing exchange, which lasts a little over a minute, raises some very interesting questions. For example, why did Ball not declare Sean's comment off limits, since it was not in line with her agenda? Why did she not intervene during the conversation between Sean and Sheena? Why did she not resolve the content issue at the end of their discussion?

The answers to these questions have to do with Ball's beliefs and practices, and with her long-term goals for instruction. Sean's comment to Sheena, although focused on content and thus not in line with Ball's short-term meta-level agenda, is very much in line with a number of Ball's beliefs and goals for her students.

Ball has worked very hard to shape her classroom so that it functions as a productive mathematical discourse community, in which students engage in mathematical sense making. Her students have learned to probe one another's ideas with respect, and to justify their own statements on mathematical grounds. Ball wants the students to see the mathematics as making sense, not to accept it because "the teacher says so." (Note learning beliefs L1 through L6 in Appendix 2C.) Of course, the teacher is a major catalyst for sense-making discussions—especially early in the year, when the students have not learned to carry on respectful mathematical conversations. However, when a conversation is both respectful and mathematically substantive, Ball will tend to step back and let it run its course. She believes that her students need to grapple with complex ideas. Giving "the answer" prematurely can deprive students of the opportunity to do sense making on their own, and perhaps even of the confidence that they can do it. She is, of course, mindful and attentive as the students conduct their conversation; she is ready to intervene if the discussion takes a turn that requires it. Her silence during the exchange between Sean and Sheena does not indicate passive withdrawal. Rather, Ball is monitoring the conversation both because intervention may become necessary and because the conversation may provide grist for later discussion or reflection.

In terms of the model, the exchange between Sean and Sheena that begins in

Lines 10 through 14 of the transcript results in the increased activation of a number of Ball's high-priority beliefs:

• L4 and L5, about the value of learning through exchanges of the type in which Sean and Sheena are engaged;

• T1, T3, and T4, about the importance of providing opportunities for students to think on their own and to express their ideas;

• S1 and S4, about students' feeling free to express themselves appropriately (Sean's comment is made using special classroom language that signals a respectful demurral: "I disagree because . . .");

• C1, C2, C3, C4, and C5, about the fact that the classroom environment should allow room for exchanges like the one taking place; and

• M1, M2, and M3, in that Sean and Sheena are engaging in real mathematics.

Because this constellation of beliefs is at a high level of activation, Goal H is set at a very high level of activation, and Ball goes into "wait and watch mode." As Ball monitors the conversation, nothing in the interaction exceeds her threshold for intervention, so she allows the conversation to run its course.

Finally, I observe that Lines 9–20C represent a second iteration of the routine outlined in Figure 2.3. Ball calls on Sean in Line 9. Sean's comment to Sheena in Line 10, diverting the conversation from Ball's reflective agenda, represents a move along the [D1] = "yes" pathway. Ball's decision to let the conversation between Sean and Sheena play out signifies a [D2] = "yes" decision. The conversation itself, in Lines 10–19, represents action A4. In Lines 20A–20C Ball brings the conversation to a gentle close. Lines 20D through 20F represent an attempt to return to her original agenda ([D5] = "yes").

Episode 2 is represented schematically in Figure 2.5. Detail is once again given in Appendix 2A.

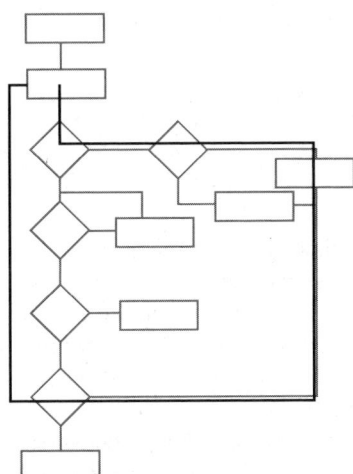

Figure 2.5. A schematic representation of the flow of lines 9–20C of the lesson transcript.

Analysis of Episode 3

Episode 3 is nearly as straightforward as a classroom episode can be. Ball re-introduces her agenda in Lines 20D–20F, calling on Mei in Line 20G (action [A2]). Mei's remark in Line 21, "I thought that zero was always going to be a even number, but from the meeting I sort of got mixed up because I heard other ideas I agree with, and now I don't know which one I should agree with," is right on target ([D3] = "no"). Ball's question in Line 22, "So what are you going to do about that?" is a request for elaboration (action [A5]). Mei's response in Line 23, "I'm going to listen more to the discussion and find out." is at the meta-level and does not require further comment; hence [D4] = "no." The exchange is now rolling smoothly, and Ball continues her debriefing agenda ([D5] = "yes") by asking, "Other people?"

The schematic representation of Episode 3 is given in Figure 2.6.

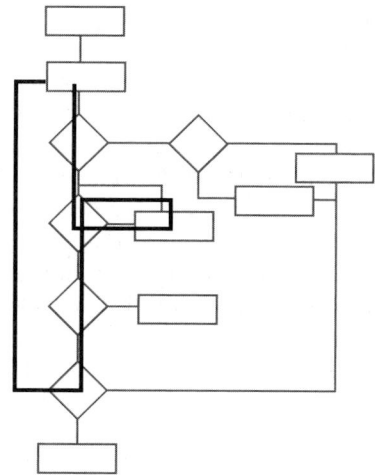

Figure 2.6. A schematic representation of the flow of lines 20D–24 of the lesson transcript.

Analysis of Episode 4

As noted in Section 3, Episode 4 represents a significant challenge for analysis. In Line 25 Nathan announces that he has "revised" his thinking, and he provides a reason for that change of mind. Although his comment is mathematical ("I think zero's special") rather than meta-mathematical (dealing with his impressions of the meeting), it is consistent with Ball's announced agenda and provides the raw material for a conversation directly in line with it.

At this point Ball deliberately deflects the conversation from her announced agenda. In Lines 26A–B ("Can I ask you a question about what you just said? And then I'll ask people for more comments about the meeting"), she announces that she is putting her agenda on hold and will return to it later. Then, over the course of Lines 26 though 43, Ball frames and asks the following question: is Nathan claiming or conjecturing that every even number is twice another even number? Clarifying Nathan's answer to this question (which, for purposes of easy reference, I shall refer to as the "evenness conjecture") occupies Ball and the class through item 58. This episode lasts for slightly more than two and a half minutes.

Asking Nathan whether he believes the evenness conjecture is a strikingly unusual decision. Ball forsakes (at least temporarily) an agenda that she has worked hard to establish. Moreover, in doing so she does not address the main mathematical content of Nathan's statement. Nathan's major claim in Line 25 is that the number zero is "special." In making that claim, he picks up the line of conversation established by the other students. Sheena had said that she thought zero was even. Sean had challenged her assertion (or at least its justification), and Mei had said she was confused about the issue. Now Nathan joins the fray. Yet Ball does not address this central mathematical point. Rather, she focuses on part of his supporting argument, the fact that that adding a pair of (the same) even numbers results in another even number.

The issue at hand is whether this unusual move on Ball's part has a reasonable explanation. Ball's decision to ask Nathan about the evenness conjecture is explicit and announced, so it cannot be dismissed as being "random." To explain Ball's teaching in this episode in terms of the model, one must show that her decision to determine the strength of Nathan's belief in the evenness conjecture is consistent with her knowledge, beliefs, and goals. Indeed, one must show that it is an entirely justifiable choice, in terms of the kind of cost-benefit analysis that lies at the core of the model's decision making. Here, in brief, is that argument.

Classroom history and her intended course of action play a fundamental role in shaping Ball's priorities. Ball's class has been exploring the properties of even and odd numbers. In an earlier class Betsy had conjectured that the sum of an odd number of odd numbers is odd, and that the sum of an even number of odd numbers is even (Ball, January 16, 1990). As noted in Section 2, Ball wanted to devote a large part of the January 19 lesson to the pursuit of those ideas.

Ball has a number of reasons for wanting to know whether Nathan and Betsy (or other students) believe the evenness conjecture to be true. She has a deep and abiding interest in making sense of her students' understandings—in seeing how they put ideas together and how ideas take shape for them. Hence she is predisposed to explore her students' understandings when circumstances permit. In addition, Ball is trying to shape her classroom community into a productive mathematical community. Mathematical communities produce results that are grounded in shared understandings. Ball is concerned with the question of whether the class members have a shared foundation for the discussions to follow. Consider, for example, this note in her January 16, 1990, class log:

> "One interesting dimension of this issue has to do with the centrality and significance of definition in a mathematical community. If Ofala believes, using her definition of even and odd numbers, that three is even, while someone else thinks

it's odd, then discussing conjectures about what happens when you add an odd number to an even number are not likely to be easy, even feasible."

Hence Ball needs to know what her students believe about the evenness conjecture if they are to be working on related conjectures later in the day. Nathan's comment in Line 25 raises for Ball the issue of the evenness conjecture and the role it has played in recent conversations. The question of who believes in the conjecture has been hanging in the air.

With this concern as background, let us consider some of Ball's options after Nathan makes his comment in Line 25. The question is, given Ball's beliefs and goals, and the options she might consider, which of those options might result as plausible choices on the basis of a cost-benefit analysis?

Here are what appear to be the three most plausible options at this point:

- *Option 1:* Let Nathan's observation pass and reorient the class to her agenda of commenting on the previous day's meeting. This, in essence, is how Ball responded to the Sean-Sheena exchange.

- *Option 2:* Pursue the central mathematical issue raised by Nathan: whether zero is even, odd, or "special."

- *Option 3:* Ask Nathan and Betsy whether they believe that the evenness conjecture is true.

The costs and benefits of pursuing Option 1 are as follows. On the plus side is lesson continuity via the continued pursuit of Ball's highest priority goal at this point, having the students reflect on the meeting. On the minus side, Ball does not get to clarify the nature of Nathan and Betsy's understandings, which would help lay the foundation for the main focus of the day's lesson. Less important but also on the minus side, the class does not get to pursue the mathematics involved in the exchange (the nature of zero).

On balance, the pluses outweigh the minuses. This choice is a plausible option, which one might well expect to be the pathway of choice.

The costs and benefits of pursuing Option 2 are as follows. On the plus side, the nature of zero is a very important issue for this mathematical community. It had surfaced the previous day and (despite Ball's attempts to focus discussion at the meta-level) had already been addressed by Sheena, Sean, Mei, and Nathan. That this issue will almost certainly be pursued by the class at some point. On the minus side, the very importance of the issue guarantees that the discussion of whether zero is even, odd, or special will be protracted. Once the class focuses its attention on that topic, other topics are almost certain to fall by the wayside. Addressing that issue now will most likely put an end to Ball's current attempts to debrief her students about the meeting.

On balance, those costs are too high. Ball is not yet prepared to let go of the debriefing agenda. This option will not be pursued.

The costs and benefits of pursuing Option 3 are as follows. On the plus side are many of the considerations mentioned above. Ball gets to clarify the nature of Nathan's and Betsy's beliefs regarding the evenness conjecture[7]. This issue is important on its own (Ball wants to understand her students' understandings) and because knowing what her students believe is essential for later classroom discussions—it establishes a firmer grounding for issues that she intends to pursue later in the lesson. On the minus side, the class does not get to pursue the mathematical issue that is obviously high on its agenda, the nature of zero. In addition, focusing on the evenness conjecture represents a potentially serious disruption of her debriefing agenda. If the students get involved in something else, can she bring them back to it? A primary consideration, then, is just how disruptive the pursuit of Option 3 might be to her debriefing agenda. Here is where Ball's move to find out what Nathan and Betsy think—but not to pursue the mathematics involved— makes a big difference. In posing a question of the type "I just want to clarify something for a second. Do you believe that X is true?" Ball demonstrates her intention to get a quick answer and then to return to her debriefing agenda.

In sum, pursuing Option 3 is likely to result in some benefits (understanding where Nathan and Betsy stand on an issue and laying the groundwork for later work) at reasonably contained costs (the short-term disruption of the debriefing agenda). Given Ball's knowledge, goals, and beliefs, this tactic is a reasonable, although not obvious, choice.

As noted earlier in this paper, I make no claim that Ball's choice of Option 3 was consciously made along the lines outlined in the foregoing—teachers hardly have the time to engage consciously in the kinds of cost-benefit analysis discussed here. One can quite reasonably assert, however, that Ball was aware of the risks of pursuing the evenness conjecture, and that she believed that exercising Option 3 instead would enable her to return to her debriefing agenda. After all, in Line 26 she signaled her priorities by announcing that their exchange would be short ("Can I ask you a question about what you just said?") and that the debriefing was still her highest priority item ("And then I'll ask people for more comments about the meeting").

In sum, Ball's decision in Line 26 is entirely consistent with a model that proceeds on a cost-benefit basis. Let us continue with a complete trace of Episode 4.

Nathan's statement in Line 25 triggers for Ball a number of issues related to her plans for the main substance of the day's lesson. Thus, it represents passage through the [D1] = "yes" decision point in Figure 2.3. Specifically in terms of the model, Nathan's statement results in the increased activation of beliefs L1, T2, C3, and C5, and of goals C and J. On that basis Ball decides to pursue Nathan's understanding of the evenness conjecture ([D2] = "yes"). In Lines 26–28 she articulates the question at hand. Starting in Line 30 (consistent with the high activation of goals C, I, and J), she brings Betsy into the conversation. She begins in Line 34 to clarify the evenness conjecture. She does so in interaction with Nathan (and with some sidebar comments by other students) in Lines 34, 39, 41, and 43.

---

[7] A very important point to note is that Ball's goal for this exchange is only to clarify for herself whether Nathan and Betsy believe that the evenness conjecture is true. She is not planning to actually discuss the truth of the conjecture at this point. Such a discussion might well be time-consuming.

In Line 45 Betsy, addressing Nathan, raises the case of six—a counterexample to the evenness conjecture. At this point the class is functioning as a mathematical community once again, discussing an issue of mathematical substance. Much as in the Sean-Sheena exchange that took place in Lines 10–14, the beliefs L4, L5, T1, T3, T4, S1, S4, C1–C5, and M1–M3 are activated, along with Goal H. Ball goes into active monitoring mode. In Line 50 Betsy completes her argument. She has shown that six, which is even, is twice an odd number (three) and not twice an even number. That fact means that the evenness conjecture cannot be true. The ensuing exchange between Nathan and Betsy (with some interleaved comments from Sean) results in Nathan's clarification (or possible revision) of his stance. Nathan says, in essence, that you can keep building larger and larger even numbers if you start with two, double it, double the result, and so on. Betsy's observation about six is taken into account in Nathan's final summation, which takes place over Lines 56 and 58: "Yeah, I'm not going by every single number. Like, two, four, six, eight." This statement refutes the evenness conjecture publicly, and brings Episode 4 to its conclusion.

Before proceeding with the analysis of Episode 5, I want to highlight two major issues raised by the events in Episode 4. First, as discussed previously, Ball's decision to do a "status check" regarding Nathan and Betsy's belief in the evenness conjecture has been shown to be an entirely plausible action that is consistent with her goals, beliefs, and knowledge. Second great consistency is evident in Ball's decisions and actions—and, I claim, in the underlying set of goals and beliefs that generate them.

In many ways, the exchange between Betsy and Nathan that takes place in Lines 45 through 58 is reminiscent of the exchange between Sean and Sheena in Lines 10 through 19. In both instances, a student challenges another student's statement. The challenge is respectful and thus within the social norms that Ball wants to encourage and maintain. The challenging student provides a solid mathematical warrant for his or her statement, so his or her comment is within the mathematical norms that Ball wants to encourage and maintain. I claim that in both exchanges, essentially the same set of beliefs regarding students, teaching, learning, and mathematics gets increased activation (see Aguirre & Speer, 2000, for a discussion of "belief bundles"). This outcome results in the same set of goals being highly activated, and thus in the selection of the same action plan—a "wait and watch" monitoring routine. Finally, in both instances, when the conversation draws to a close on its own, Ball returns the discussion to her debriefing agenda.

The schematic representation of Episode 4 is given in Figure 2.7. Detail is once again given in Appendix 2A.

Analysis of Episode 5

In Line 59 Ball once again invites students to comment on the previous day's class (step A1 of the routine in Figure 2.3). Sean responds in Line 60. Sean was beginning to explore out loud the idea that the number six is both even and odd.

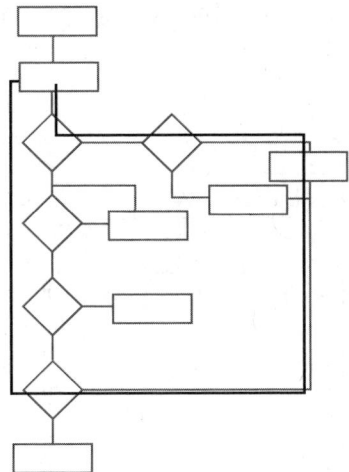

Figure 2.7. A schematic representation of the flow of lines 25–58 of the lesson transcript.

His idea was that six is even because it is made up of two sets that are the same size (two threes), and that it is odd because it is made up of an odd number of sets that are the same size (three twos).[8] However, Sean's statement was ambiguous, and Ball did not hear it that way. Ball writes that she thought Sean's comment was connected with the discussion that had just come to a close. "I thought that Sean's point was that two odd numbers could also make an even number. I assumed that we could now move on with our discussion of the meeting and then move from there to what we were supposed to be working on today" (Ball, undated, page 1).

That is, Ball heard Sean's comment about two threes making six in the context of the previous exchange over the evenness conjecture. With this understanding she confirmed with Sean that six is twice an odd number (Line 63) and then (Line 65) that this statement meshed with Nathan's: some but not all even numbers are made up of a pair of even numbers.

Sean's comments in Lines 60–62 and Ball's decision in Line 63 to clarify Sean's statement and relate it to the previous conversation represent passage through the [D1] = "yes" pathway (as he says himself, Sean's comment is not about the meeting) and then the [D2] = "no" pathway in Figure 2.3. Her subsequent request for comments represents passage through the [D5] = "yes" pathway. Episode 5 thus

---

[8] Sean's idea may seem strange to readers who are not familiar with the story of this day's lesson—after all, we know that any whole number must be either even or odd. But for a student who is just starting to make sense of odd and even numbers, that point is not necessarily clear. For example, one can define squares and cubes. Some squares, such as 64, are also cubes. Could something similar not happen with even and odd numbers? A student might know that a number such as six is even but wonder why it could not be odd too. At this point the class has developed a clear working definition of even: even numbers are "made up" of two numbers that are the same size. The class does not have a firm working definition of odd. Sean is exploring the idea that an odd number is something made up of three numbers that are the same size.

represents a fifth tour of Figure 2.3. This path is represented in schematic form in Figure 2.8.

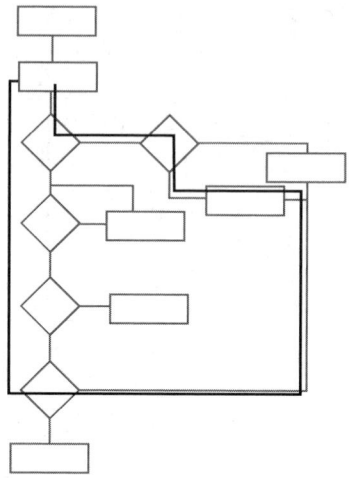

Figure 2.8. A schematic representation of the flow of lines 59–67 of the lesson transcript.

From a description to a model

As discussed in Section 2.2, a representation of a lesson like the one given in Appendix 2A can be read in two ways. Reading from left to right, one has the kernel of a descriptive story such as the one told in the preceding section. The lesson segment as it unfolds is parsed iteratively into smaller episodes, and the goals and beliefs that are highly activated during each episode are identified. If, however, one reads Appendix 2A from the middle on—identifying the knowledge, goals, and beliefs that are at high levels of activation, then implementing a decision-making mechanism that selects a course of action to achieve the high-priority goals—one then has the kernel of a model. In this section I work through Appendix 2A in that way, showing how it can be used to model Ball's teaching.

The fundamental shift between the previous section and this one is as follows. In the previous section I made the following claim: "Here is what Ball did, and here is a plausible explanation of why she did it." In this section I make the following claim: "Here is a model whose initial state consists of a set of beliefs, goals, and knowledge that I believe represents (some of) Ball's beliefs, goals, and knowledge when she taught the lesson on January 19, 1990. The model makes its decisions on a cost-benefit basis, as previously described. When one works through the model, it produces decisions that are remarkably consistent with those made by Ball." Entailments of this claim are explored in the "Discussion" section.

I shall now indicate how the model represented in Appendix 2A would simulate Ball's decision making. For ease of reference, I shall call the model MB.

Before the simulation begins, MB is "primed" with the beliefs and goals listed in Appendices 2C and 2B, respectively. Corresponding to those beliefs and goals (specifically the very high activation level of goals D and E and the relatively high activation of beliefs L3, L4, L6, T1–T4, C5, S1, and S2), MB selects and begins to implement the routine in Figure 2.3. This action results in Line 1 of the transcript, in which MB sets the stage for the discussion and calls on a student.

Sheena's response, in Line 2, is in the general ballpark ([D1] = "no") but vague ([D2] = "yes"), thereby causing MB to implement its "clarify through scaffolding" routine, which is implemented in Lines 3 (ask for clarification) and 5 (ask for an example). Sheena's comment in Line 6 is deemed appropriate for abstraction ([D4] = "yes"), so MB does so in Line 7. Given that the debriefing discussion has just begun, [D2] = "yes," MB embarks on the next iteration of the routine in Line 9.

Sean's comment to Sheena in Line 10 addresses a content issue rather than the reflective agenda ([D1] = "yes"). It is, however, respectful and mathematically substantive. The exchange triggers the increased activation of beliefs L4, L5, T1, T3, T4, S1, S4, C1-C5, and M1–M3; thus [D2] = "yes." Goals D and E remain active, but lowered in priority for the time being, and Goal H is very strongly activated. As a result, MB implements a "monitor with the possibility of intervention" routine, which remains in place through the end of Line 19. When the exchange between Sean and Sheena is completed, MB makes a closing comment in Lines 20A–C. At that point, Goal H is deactivated and the set of beliefs and goals that were active at Line 1 are once again given high levels of activation. [D5] = "yes," and in Line 20D MB begins another iteration of the routine.

Mei's response in Line 20 is in line with MB's agenda ([D1] = "no"). Her comment about being confused ([D3] = "yes") raises a natural question, which MB asks in Line 22. Mei's response in Line 23 is crystal clear, needing no further elaboration ([D4] = "no"). The reflective agenda is going well, so [D5] = "yes"; MB, in Line 24, asks for more comments. Once again, the beliefs and goals that accompany the start of the routine (goals D and E and beliefs L3, L4, L6, T1–T4, C5, S1, and S2) are given strong activation.

Nathan's response in Line 25 boosts the activation levels of beliefs L1, T2, C3, and C5; concomitantly, Goal J receives very strong activation. Here at least three alternative pathways through Figure 2.3 are plausible. A central part of Nathan's assertion concerns the "special" nature of 0, for which he gives a plausible explanation. This statement might be considered potentially "on target" ([D1] = "no"), and MB might ask a clarifying question aimed at meta-level issues—for example, "was there something in yesterday's conversation that led you to think this way?" Alternatively, since Nathan's comment focuses on content instead of addressing Ball's meta-level agenda, it might be considered off-target ([D1] = "yes"). MB might decide that the content issue should not be pursued ([D2] = "no") and consequently would seek closure (step [A4]) and begin the routine once again. Or, given the very strong activation of Goal J, the activation of goals B and C, and the assessment that the debriefing agenda could withstand a short-term detour, MB might ask Nathan for clarification regarding his view of the evenness conjecture. All three of these pathways are plausible, and none of them is overwhelmingly strong in terms of activation or cost-benefit analysis. Any of these might be chosen in one run of the model. (In multiple runs, each would be chosen some of the time.)

If MB chose to pursue either of the first two options mentioned in the foregoing—leading Nathan to a meta-level conversation or deciding not to pursue the mathematical content of his remark—one could imagine running out the tour of Figure 2.3 in straightforward ways. If, as Ball did, MB decides to pursue Nathan's understanding of the evenness conjecture, MB's question takes us through Line 28. Betsy had made a conjecture similar to the evenness conjecture, and the success of discussions planned for later in the lesson might depend on whether Nathan, Betsy, and perhaps some other students believe the conjecture to be true. Thus goals I and J are activated. MB acts to bring Betsy into the conversation in Lines 3–32, and in Lines 34–43 to clarify the issue for all the students (cf. Goals G and J).

The context of discussion has shifted to the evenness conjecture. Betsy's comment in Line 45 is right on target: it challenges Nathan respectfully and on mathematical grounds. Beliefs L4, L5, T1, T3, T4, S1, S4, C1–C5, and M1–M3 are activated. Goal J remains activated, and Goal H is given a very strong level of activation. MB once again implements a "monitor with the possibility of intervention" routine, which remains in place through the end of Line 58.

MB then returns to step A2 of the routine, asking in Line 59 for more comments about the meeting. Sean's comment is not about the debriefing agenda; hence [D1] = "yes." How MB responds to Sean's comment in Line 60 depends on what understandings it has. If MB is given Ball's understanding at the moment, that Sean is making a comment related to the discussion of the evenness conjecture, then [D2] = "no:" the topic has been discussed at adequate length and Sean's comment does not add to what has been said. MB notes that Sean's statement (six is made up of two threes, which are odd) is consistent with Nathan's comment that some but not all even numbers are made up of a pair of the same even numbers. This comment addresses (MB's understanding of) Sean's comment. With Sean's comment addressed, MB decides to continue the debriefing agenda ([D5] = "yes) and asks for more comments in Line 67. Thus ends one virtual run of MB.

Having worked through the model, we need to compare MB's decisions with those made by Ball. When one examines the two sets of decisions, the correspondence between the model's and Ball's decisions is questionable in two places. The first is in response to Nathan's comment in Line 25. As indicated previously, three paths of action are plausible. MB's cost-benefit analysis does not does settle clearly on any of them. Hence MB's decision making does not necessarily replicate what Ball actually did. A problem does not necessarily exist here. One might reasonably judge that if Ball had faced the same situation on another day, she might have acted differently—she had no compelling reason to make the choice she did. In essence, the model says the same thing.

The second is MB's response to Sean's comment in Line 60. Selecting an action in line with Ball's actual choice depends on a (mis)understanding of what Sean was saying, consistent with Ball's understanding of it. But that dependence is precisely as it should be: MB can be expected to make decisions like Ball's only if it has Ball's knowledge (whether that knowledge is accurate in an "objective" sense or not), goals, and beliefs.

Overall, then, MB's decision making is consistent with Ball's. Some of the implications of this concordance are discussed in the next section of this chapter.

DISCUSSION

This discussion has two components, a look at the "dilemmas/decisions" distinction and a commentary on the modeling enterprise.

First, dilemmas and decisions. Metaphorical systems have great power, and the term dilemmas has a humanist ring to it. When we hear the term , we envision teachers being confronted with difficult choices—teachers working to resolve the inevitable tensions that result from trying to achieve many things and honor many constraints at once. In contrast, the terms *models* and *decision making* have rather reductive and dehumanizing connotations. Think, for example, of the image you conjure up when envisioning a teacher who "follows" the decision tree in Figure 2.3.

The research presented in this chapter indicates that a conflict need not exist between the two characterizations. Teaching dilemmas are real, and they are felt that way: teachers do what they can to honor multiple and often conflicting goals. Many of the situations described in this chapter are of that type—at numerous points in the lesson segment, a decision by Ball to pursue one option effectively foreclosed other options. Ball's decisions were made spontaneously, amidst the flow of classroom events. She had no time for the careful consideration of options. Ball had to do what she could, on the fly, to honor her commitments and values.

One of the things I find most interesting about the analysis given in this chapter is that it reveals a remarkable consistency underlying Ball's teaching. Ball is extremely flexible and responsive both to her students and to circumstances as they evolve in her classroom. She is, as noted above, attending to multiple and oftentimes conflicting goals. Yet her teaching decisions can be modeled with great fidelity by a reasonably straightforward decision procedure.

Again, I am not suggesting that Ball follows this decision procedure either consciously or unconsciously. However, the fact that her teaching can be modeled suggests that it involves greater systematicity than appears on the surface—that it reflects, at a deep level, a significant amount of coherence in the way that Ball honors her commitments and values. Here is where the nature of the modeling process makes a difference. Models of the type described in this chapter are more than abstractions of regularities of behavior. The decision-making mechanism in the model calls for identifying potential decisions and for sorting through them. The decisions are made by cost-benefit analysis on the basis of the (attributed) goals, knowledge, and beliefs of the teacher. This analytic enterprise points to the sources of the decisions teachers make as they teach.

The fact that a teacher's beliefs, goals, and knowledge play such an important role in the teacher's decision making has interesting implications for professional development. It supports an argument that having teachers reflect on their decision making—thinking about the options they might have considered and did not, about why they made the choices they did, and about how those choices reflect their beliefs, goals, and knowledge—can serve as a catalyst for professional growth.

I conclude with some notes on the status and utility of the modeling enterprise. The analysis of Ball's lesson represented a major test case for the theory of teaching-in-context. A fundamental question regarding any such theory is its scope, that is, the range of events to which the theory can be said to apply. As discussed under "Aspects of a Theory of Teaching in Context," Berkeley's Teacher Model Group had succeeded in modeling a collection of lessons that could be seen as covering a significant amount of territory: beginning and experienced teachers conducting both traditional and innovative lessons. However, the lessons were all at the secondary level or above, and all were in mathematics and science classes. They were also driven by the teacher's agendas. Ball's lesson is quite something else.

Issues of scope were addressed in the first major paper that described the theory (Schoenfeld, 1998b). The TMG had completed the analyses of lessons taught by Minstrell, Nelson, and Schoenfeld and had begun to work on Ball's January 19, 1990, lesson. We noted that we had not yet been successful in our attempts to model the lesson and that our failure (if we did fail) would serve to establish the boundaries of what we were capable of modeling. In a commentary on our work, Borko and Peressini (1998) explored possible reasons for our difficulties. They wrote,

> We appreciate the candidness with which Schoenfeld presented the problems his group encountered as they tried to analyze and model the Sean Numbers lesson. In this section we further explore some of the concerns he identified. Based on this exploration, we suggest that Ball's conception of teaching and the classroom mathematics pedagogy she enacts may be incompatible with the assumptions that underlie Schoenfeld's theory. (Borko & Peressini, 1998, p. 99)

Borko and Peressini's argument, grounded in descriptions of teaching set forth by Lampert (1985) and Ball (1993, 1997), is that the very conception of a teacher as a "dilemma manager"—someone struggling to balance conflicting goals, often improvising as he or she "juggles competing priorities and chooses particular strategies in particular situations, not because they solve problems but because they temporarily resolve conflicts in a way that makes sense for the moment" (Borko & Peressini, 1998, p. 99)—may be fundamentally incompatible with the rationalist approach that underlies the modeling enterprise.

Likewise, Leinhardt (1998) voices reservations about scope:

> Consider the examples and claims from Schoenfeld's piece. Does it matter that all the teachers who were studied successfully are men? Probably not. Does it matter that the sample comes from high school and above? Probably yes. Does it matter that the content is mathematics/physics? Probably yes. (Leinhardt, 1998, p. 127)

This latter point is crucially important to Leinhardt, who emphasizes that domain epistemology makes a big difference. The very nature of history instruction must differ from the nature of mathematics instruction, she argues, because the essential ideas to be learned are fundamentally different. Leinhardt also emphasizes that teaching, especially at the elementary level, is fundamentally interactive and dynamic:

> By their own report, teachers are constantly thinking about and responding to the specific individuals in their classes, the clusters of students in their classes, and the ghosts of students who populated former classes. Teachers enter into implicit and explicit conversations with these students. . . .Teachers may anticipate and avoid student responses, they may anticipate and build off student ideas, but they are aware of and actively respond to the specific students in the class. More importantly, teachers are juggling their own cumulative understandings against their goals and actions. (Leinhardt, 1998, p. 131)

These comments underscore the importance of the Ball analysis as a test case for the TMG's work. First, Borko and Peressini suggest that the conception of "teacher as dilemma manager" is fundamentally incompatible with the notion of teaching as a modelable enterprise. The analysis in this chapter challenges that assumption. Formal representations of modelable behavior (e.g., the representations in Figure 2.3 or Appendix 2A) may seem cut-and-dried, and their form may tend to obscure the tensions and dilemmas over which teachers agonize. I believe, however, that the modeling approach described here provides tools that help elucidate those dilemmas and the ways that teachers deal with them. Moreover, I believe that reflecting on the knowledge, goals, and beliefs that underlie the decision-making process will prove to be an important source of professional development for teachers.

Does domain epistemology matter, as Leinhardt claims? Undoubtedly it does, in terms of what teachers do. Whether it limits what can be modeled remains to be seen. Do the voices of current and even past students matter? Undoubtedly they do. I hope the analysis in this chapter, in suggesting how Ball's decisions are made (see, for example, the discussion of how events prior to the lesson in question shaped Ball's response to Nathan's statement in Line 26), shows that models of the type developed here can take those voices and their echoes into account.

I conclude by commenting on a specific and rather surprising finding from the current analysis.

Two of the case studies discussed in Schoenfeld (1998b) were a high school physics lesson taught by Jim Minstrell and a college mathematics problem-solving lesson taught by myself. On the surface—in terms of teaching style and lesson content—the classes seem to be very different. However, the modeling process pointed to similarities in the ways the two teachers solicited comments from students and orchestrated discussions around them. Both Minstrell's interactions with students and mine were modeled with relatively simple decision procedures, which turned out to be similar in structure and pointed to previously unseen parallels in our teaching.

When TMG began the analysis of Ball's lesson, it seemed a world apart from the other two. As noted previously, the first 6 minutes of the January 19, 1990, episode contain many unexpected twists and turns. The deep structure of that lesson segment remained inaccessible to us for quite some time. Only after we developed the representation of the routine given in Figure 2.3 did we realize that it could also be used to model Minstrell's and my lessons.

The parallels are surprising. Side-by-side, videos of Ball's, Minstrell's, and my classrooms look very, very different; they reveal differences in students, in content, and in classroom style and dynamics. Ball was asking her students to reflect on a meeting held the previous day; Minstrell and his students were trying to determine the "best number" to represent a collection of data; my students and I were working on a complex geometry problem. At the same time, Ball, Minstrell, and I are all known for having developed a particular kind of instruction—instruction that focuses on developing meaning and understanding, and in which classroom discussions are highly interactive. Ball, Minstrell, and I developed our instruction independently. However, the three of us share certain epistemological assumptions—and those assumptions have pedagogical consequences. All of us work to "surface" students' ideas, and to have those ideas serve as the base for reasoned classroom discourse. A truly fascinating outcome is that models of our teaching bypass the surface differences between us and point to deep and underlying similarities in our classroom actions and decisions. That result suggests that the kind of routine described in Figure 2.3 may be somewhat general, and that studying such a routine explicitly may be useful for teachers who want to develop highly interactive classrooms that focus on developing meaning and understanding. (For a detailed exploration of this conjecture, see Schoenfeld, 2002.)

Acknowledgments

Although this chapter has a single author, the work is the product of a truly collaborative enterprise, the Teacher Model Group at Berkeley. Members of the group during the genesis of the paper included Julia Aguirre, Claire Bove, Ming Ho, Lani Horn, Andrew Izsak, Krystal Jones, Michael Keynes, Brett Lorie, Rodrigo Madanes, Sue Magidson, Antonio Olmedo, Tamar Posner, Manya Raman, Ann Ryu, Rachel Sacher, Alex Shefler, Natasha Speer, Joe Wagner, Rachel Westlake, and Dan Zimmerlin. Thanks to Julia Aguirre, Abraham Arcavi, Andrew Izsak, Nicki Kersting, Cathy Kessel, Sue Magidson, Ann Ryu, Natasha Speer, Joe Wagner and Dan Zimmerlin for their comments on draft versions of this manuscript. Special thanks go to Deborah Ball for her willingness to share her teaching data and her openness in discussing a wide range of issues related to her teaching.

REFERENCES

Arcavi, A., Kessel, C., Meira, L., & Smith, J. P. (1998). Teaching mathematical problem solving: An analysis of an emergent classroom community. In A. H. Schoenfeld, J. J. Kaput, & E. Dubinsky (Eds.), *Research in collegiate mathematics education (Vol. III)* (pp. 1–70). Washington, DC: Conference Board of the Mathematical Sciences.

Aguirre, J., & Speer, N. (2000). Examining the relationship between beliefs and goals in teacher practice. *Journal of Mathematical Behavior, 18*(3), 327–356.

Ball, D. L. (Undated). Annotated transcript of segments of Deborah Ball's January 19, 1990, class. Distributed by Ball at the research presession to the 1997 annual meeting of the National Council of Teachers of Mathematics, San Diego.

Ball, D. L. (January 16, 1990). Teacher's class log.

Ball, D. L. (1993). With an eye on the mathematical horizon: Dilemmas of teaching elementary school mathematics. *The Elementary School Journal, 93*(4), 373–397.

Ball, D. L. (1997). What do students know? Facing challenges of distance, context, and desire in trying to hear children. In B. J. Biddle et al., (Eds.), *International handbook of teachers and teaching* (pp. 769–818). New York: Macmillan.

Ball, D. L. (April 4, 1999) Personal communication.

Ball, D. L., & Lampert, M. (1999). Situating research within an education context: A case study in mathematics teaching and learning. In E. R. Lagemann & L. Shulman (Eds.), *Issues in education research: Problems and possibilities* (pp. 371–398). San Francisco, CA: Jossey-Bass.

Ball, D. L., & Wilson, S. J. (1996). Integrity in teaching: Recognizing the fusion of the moral and the intellectual. *American Educational Research Journal, 33*(1), 155–192.

Borko, H., & Peressini, D. (1998). Commentary on "Toward a theory of teaching-in-context." *Issues in Education, 4*(1), 95–104.

Lampert, M. (1985). How do teachers manage to teach? Perspectives on problems in practice. *Harvard Educational Review, 55*(2), 178–194.

Lampert, M. (2001). *Teaching problems and the problem of teaching.* New Haven: Yale University Press.

Leinhardt, G. (1998). On the messiness of overlapping goals and real settings. *Issues in Education, 4*(1), 125–132.

Schoenfeld, A. H. (1998a). Reflections on a course in mathematical problem solving. In Alan H. Schoenfeld, James J. Kaput, & Ed Dubinsky (Eds.), *Research in collegiate mathematics education (Vol. III)* (pp. 81–113). Washington, DC: Conference Board of the Mathematical Sciences.

Schoenfeld, A. H. (1998b). Toward a theory of teaching-in-context. *Issues in Education, 4*(1), 1–94.

Schoenfeld, A. H. (1999). Models of the teaching process. *Journal of Mathematical Behavior, 18*(3), 243–261.

Schoenfeld, H. (2002). A highly interactive discourse structure. In J. Brophy (Ed.), *Social constructivist teaching: Its affordances and constraints* (Vol. 9 of the series Advances in Research on Teaching), pp. 131–170. Amsterdam: JAI Press.

Schoenfeld, A. H., Minstrell, J., & van Zee, E. (1999). The detailed analysis of an established teacher's non-traditional lesson. *Journal of Mathematical Behavior, 18*(3), 281–325.

van Zee, E., & Minstrell, J. A. The promise of a theory of teaching-in-context. *Issues in Education, 4*(1), 141–147.

Zimmerlin, D., & Nelson, M. (1999). The detailed analysis of a beginning teacher carrying out a traditional lesson. *Journal of Mathematical Behavior, 18*(3), 263–279.

APPENDIX 2A

Ball Lesson Transcript, Parsing, and Teacher Goals
(See pages 86–94)

APPENDIX 2A

Ball Lesson Transcript, Parsing, and Teacher Goals

Line	Who	Transcript	First Level of Parsing	Second Level of Parsing	Goals:

Goals: See Appendix B for the Goal Legend. Goals A, B, and C are overarching goals. Goals D and E are major content goals. Goals F,G,H,I, and J are local goals

Goal columns: A B C D E F G H I J

Line	Who	Transcript	First Level of Parsing	Second Level of Parsing
1	Ball	[Preliminaries] [A] Okay. A few delays, but I think we're ready to start now.	[Preliminaries] Getting the class started	
		[1] [B] I'd like to open, open the discussion today with um – I have a few questions about the meeting yesterday that I'd like to ask. [C] So, to begin with, I would just like everybody to put pens down, there's nothing to take notes about or do right now. [D] But I'd like you to be thinking back to yesterday to the meeting that we had on even and odd numbers and zero. [E] And I have a few questions. First – my first question is, I'd just like to hear some comments about what you thought about the meeting, what you noticed about the meeting, what you learned at the meeting, just what kinds of comments you have about yesterday's meeting? [F] And could you listen to one another's comments, so that we can um, benefit from what other people say? See what y– what you think about other people's comments? [G] Sheena, do you want to start?	[1.1] Initiation of debriefing routine, and (as planned) execution of routine through one full cycle. See sections 4.2 and 4.3 for prose descriptions of this routine. (Decisions listed in this column are represented in Figure 3.)	[1.1.1] Establishing Context and background (Action A1) Classroom management [1.1.2] Posing the main question (Action A2) Classroom management Calling on student
2	Sheena	I– I– I liked it because, well, I like talking to other classes and, and when you talk to other classes sometimes it helps.	[D1] = no [D3] = yes	[1.1.3] Student Response
3	Ball	In what way?		[1.1.4] Request Clarification (Action A5)
4	Sheena	It helps you to understand a little bit more.	[D3] = yes	[1.1.5] Student Response
5	Ball	Was there an example of something yesterday that you understood a little bit more during the meeting?		[1.1.6] Request Clarification (Action A5)
6	Sheena	Well, I didn't think that zero was – zero, um – even or odd until yesterday they said that it could be even because of the ones on each side is odd, so that couldn't be odd. So that helped me understand it.	[D4] = yes	[1.1.7] Student Response
7	Ball	Hmm. So y– So you thought about something that came up in the meeting that you hadn't thought about before? Okay.		[1.1.8] Abstract and reframe student comment (Action A6)
8	Sheena	(Nods in agreement.)	[D5] = yes	(Cycle ends)
9	Ball	Other people's comments? Sean?	[1.2] 2nd cycle of routine	[1.2.1] initiation of 2nd cycle (Action A2)

APPENDIX 2A (*continued*)

Beliefs:

See Appendix C for the Belief Legend.
Beliefs L1-L6 are about learning
Beliefs T1-T4 are about teaching
Beliefs S1-S4 are about students
Beliefs C1-C5 are about classroom environments
Beliefs M1-M3 are about mathematics

L1	L2	L3	L4	L5	L6	T1	T2	T3	T4	S1	S2	S3	S4	C1	C2	C3	C4	C5	M1	M2	M3

Relevant Knowledge

Previous days' conversations
Experience of previous year's
  conversations
Individual students' backgrounds
Her thoughts about student
understandings of the topic
and about the relevant mathematics
(evenness, oddness, the possibly
"special" nature of 0).

Ball has ready access to the
debriefing routine - it's a
standard part of her repertoire.

Decision-Making

Ball begins the class with
the intention of "debriefing"
the students about the previous
day's meeting.

She plans to implement
the "flexible, interruptable"
debriefing routine described
in Figure 3.

As described in the narrative,
Ball has a general scaffolding
  routine that she uses to
get students to clarify/
elaborate their thoughts.
This, combined with her
specific knowledge of Sheena,
suggests how much scaffolding
is necessary.

Sheena's comment is on
target ([D1] = no), so Ball
will pursue it. It is vague
(D[D3]=yes), so Ball will
provide scaffolding to help
her clarify what she means
to say.

[D3] = yes.
Sheena's statement is still
vague, so Ball will make a
2nd move for clarification,
using a standard "give me
an example" prompt in
response to a vague statement.

Part of Ball's knowledge is the
ability to abstract the content
of conversations with students
in the service of her goals, and to
present that content at an
appropriate level to her students

This time Sheena's answer
is clear ([D3] = no)
so Ball moves on.

[D4] = yes. Ball
reframes what Sheena
has said at a meta-level:
one hears new things at
meetings.

[D5] = yes
Ball pursues the debriefing
routine.

APPENDIX 2A (*continued*)

Line	Who	Transcript	First Level of Parsing	Second Level of Parsing	Goals:
					See Appendix B for the Goal Legend. Goals A, B, and C are overarching goals Goals D and E are major content goals Goals F,G,H,I, and J are local goals
10	Sean	Um, I – I – I just want to say something to Sheena, when sh– what she said about um that, that one, um– zero has to be an odd, an even number bec– I disagree because, um, because what what two things can you put together to make it?	Sean's comment raises other issues: [D1] = yes Ball decides this should be pursued:	[1.2.2] Student Response	
11	Sheena	Could you repeat what you said, please?	[D2] = yes	[1.2.3]	
12	Ball	(Aside to Betsy asking her to listen to the Sean/Sheena exchange)	Consistent with the very high activation level	Classroom Management	
13	Sean	Okay, um, I disagree with you because, um, if it was an even number, how– what two things could make it?	of goal H, Ball implements a "wait and watch" routine.	Ball is monitoring for possible intervention.	
14	Sheena	Well, I could show you it. (Moves toward the chalkboard and points to the number line above the chalkboard.) Um, I forgot what his name was–but yesterday he said that this one (points to the 1 on the number line) and each– this one is odd and this one (point to negative 1 on the number line) is odd, so this one (points to 0 between them) has to be even.			
15	Sean	But, that doesn't mean it always is even.			
16	Sheena	It could be even.			
17	Sean	It could be, but . . .		When the interaction comes to an end, Ball is ready to move on.	
18	Sheena	I'm not saying that is has to be even. I meant that it could be.			
19	Sean	You said it was. (Long pause, in which tension defuses)			
20	Ball	[A] Before we take this up again, I underst– I– I understand that this is still a problem and that we didn't a– we didn't settle it, we're probably not going to settle it. [B] Um, there's a lot of disagreement about this issue, right? [C] And you saw that the fourth graders who have been thinking about this for a long time also disagree about it, don't they? [D] I'm still kind of interested um, in hearing some more comments about the meeting itself. [E] Sheena commented that it was good to have the two classes together because she heard an idea that she hadn't thought about and it made her think about and even revise her own idea when she was in the meeting yesterday. [F] What other comments do other people have about the meeting and what happened yesterday? [G] Mei, do you have a comment?	[D5] = yes [1.3] 3rd cycle of routine	[1.2.4] Ball moves to ease the tension, emphasize the idea that sorting out ideas takes time, and segue back to her agenda of debriefing. She re-establishes the context by recapping the previous discussion [1.3.1] Ask for Comments (Action A2)	
21	Mei	Um, I h– I thought that zero was always going to be a even number, but from the meeting I sort of got mixed up because I heard other ideas I agree with and now I don't know which one I should agree with.	[D1] = no [D3] = yes	[1.3.2] Student Response	
22	Ball	Um-hm. So what are you going to do about that?		[1.3.3] Request amplification (Action A5)	

APPENDIX 2A (*continued*)

Beliefs:

See Appendix C for the Belief Legend.
Beliefs L1-L6 are about learning
Beliefs T1-T4 are about teaching
Beliefs S1-S4 are about students
Beliefs C1-C5 are about classroom environments
Beliefs M1-M3 are about mathematics

Relevant Knowledge

Decision-Making

The issue here was whether or not to intervene.

As described in the narrative, productive and respectful student exchanges are a major goal for Ball's classroom. Beliefs related to such exchanges are triggered, and a "wait and watch" monitoring routine is activated.

Mathematical content related to previous days' conversations; histories of Sean, Sheena

As is her wont, Ball lets the Sean-Sheena exchange wind down on its own. She then moves to bring closure to the exchange. She re-frames the exchange somewhat, using it to re-emphasize the idea that some mathematical ideas take a long time (and hard work) to sort out.

[D5] = yes: return to the debriefing routine.

Knowledge of Mei:
Ball can stretch Mei with a general question because she knows Mei can handle it.

APPENDIX 2A (*continued*)

Line	Who	Transcript	First Level of Parsing	Second Level of Parsing	Goals:
					See Appendix B for the Goal Legend. Goals A, B, and C are overarching goals Goals D and E are major content goals Goals F,G,H,I, and J are local goals
23	Mei	Um, I'm going to listen more to the discussion and find out.	[D4] = no [D5] = yes	[1.3.4] Student Response	
24	Ball	Other people? Nathan?	[1.4] 4th cycle of routine	[1.4.1] initiation of 4th cycle (Action A2)	
25	Nathan	Um, first I said that um, zero was even but then I guess I revised so that zero, I think, is special because um, I – um, even numbers, like they they make even numbers; like two, um, two makes four, and four is an even number; and four makes eight; eight is an even number; and um, like that. And, and go on like that and like one plus one and go on adding the same numbers with the same numbers. And so I, I think zero's special.	See the narrative in Section 4.3 for detail. Nathan's comment raises other issues: [D1] = yes Ball decides this should be pursued [D2] = yes	[1.4.2] Student Response	
26	Ball	[A] Can I ask you a question about what you just said? [B] And then I'll ask people for more comments about the meeting. [C] Were you saying that when you put even numbers together, you get another even number –		[1.4.3] This long section, in which Ball speaks slowly and in which many students comment, consists essentially of Ball asking one question: Do you believe the "evenness conjecture"?	
27	Nathan	Yeah.			
28	Ball	or were you saying that all even numbers are made up of even numbers?			
29	Nathan	Yes, they are			
30	Ball	Betsy, you said something like that yesterday, too.		Bringing Betsy into the conversation	
31	Betsy	What.			
32	Ball	Were you – were you not listening to this just now?			
33	Betsy	No.			
34	Ball	Nathan said a minute ago that when you put even numbers together you get an even number.			
35	Betsy	Mm-hm.			
36	Ball	But he also said, I think, that all even numbers are made up of other even numbers.			
37	Mei	I disagree.			
38	Sheena	(says something to Mei)			
39	Ball	Two even numbers just the same.			
40	Nathan	Unh-uh.			
41	Ball	The same even number?			
42	Nathan	Yeah, like four.			
43	Ball:	[A] Like eight is four plus four? [B] Are all the even numbers – can you do that with all the even numbers? That they'd be made up of two identical even numbers?			
44	Sean	Not– not– not–			

APPENDIX 2A (*continued*)

Beliefs:

See Appendix C for the Belief Legend.
Beliefs L1-L6 are about learning
Beliefs T1-T4 are about teaching
Beliefs S1-S4 are about students
Beliefs C1-C5 are about classroom environments
Beliefs M1-M3 are about mathematics

Relevant Knowledge

Decision-Making

[D3] = yes; [D4] = no.
The answer given by Mei
is perfect, requiring no
additional comment from the
teacher.
[D5] = yes,
and thus Ball continues
implementation of the
debriefing routine.

A great deal of relevant
knowledge is triggered by
Nathan's comment. Specifically,
her knowledge of where she
wants the lesson to go
(exploring students' conjectures
about sums of even and odd
numbers), and of particular
particular students' histories with
these conectures (e.g., Nathan's
and Betsy's) is activated.

Please see Sections 4.3 and 4.4
of the narrative for a detailed
explanation of Ball's decision
to ask Nathan about whether
he believes that every even
number is twice some other
even number. The argument,
in brief, is that Ball determines
that she can clarify Nathan's
understanding (as well as
Betsy's) in short order,
without sacrificing her
debriefing agenda. This
will enable her to pursue the
debriefing (Goals D and E)
after developing a better
undestanding of Nathan
and Betsy's grasp of
an important mathematical
idea (Goals B and C) that
will be pursued later in the
lesson (Goal J). In addition,
this allows Ball to rope
Betsy into the
conversation (Goal I).

Ball finishes posing her question
to Nathan.

APPENDIX 2A (*continued*)

Line	Who	Transcript	First Level of Parsing	Second Level of Parsing	Goals:
					See Appendix B for the Goal Legend. Goals A, B, and C are overarching goals. Goals D and E are major content goals. Goals F,G,H,I, and J are local goals
45	Betsy	(looking toward Nathan) You can't. Like six. Six is two, two, ... Six you can't get two.		[1.4.4]	
46	Sean	Six is two odd numbers to make an even, to make an even number.		Monitoring for Possible Intervention	
47	Mei	Three and three –			
48	Betsy	(still looking toward Nathan). You need three twos to make six. You can't put a four and a four or a . . .		Betsy's interaction with Nathan is in some ways parallel to the Sean-Sheena exchange in [1.2]. Once Betsy and Nathan begin to interact, Ball holds back to see what happens.	
49	Sean	Three twos???			
50	Betsy	(looking toward Nathan) Three's – Three is odd.			
51	Sean	Or, um –			
52	Nathan	I know that, but um, um I'm talking about like two plus two is four, and four plus four is eight and I just skipped the six so I just added the ones that, that add. Like the two plus two is four, and four is an even number and I'm just talking about the things that um, like –			
53	Sean	Six can be an odd number.			
54	Nathan	What I just said– the um, like two is plus two is four and four plus four is eight and um –			
55	Betsy	So what you're doing is you're going by twos and then what two equals from then you go from – all the way up .			
56	Nathan	Yeah, I'm not going by every single number. Like,	A5 (closure) achieved	This statement by Nathan asnwers Ball's question - he is not claiming the evenness conjecture.	
57	Betsy	Okay.			
58	Nathan	two, four, six, eight.	[D5] = yes		
59	Ball	[A] More comments about the meeting? [B] I'd really like to hear from as many people as possible what comments you had or reactions you had to being in that meeting yesterday. [C] Sean?	[1.5] 5th cycle of routine	[1.5.1] initiation of 5th cycle (Action A2)	
60	Sean	Um, I don't have anything about the meeting yesterday, but I was just thinking about six, that it's a. . . I'm just thinking. I'm just thinking it can be an odd number, too, 'cause there could be two, four, six, and two, three twos, that'd make six ...	[D1] = yes	[1.5.2] Student Response	
61	Ball	Uh-huh . . .			
62	Sean	And two threes, that it could be an odd and an even number. Both. Three things to make it and there could be two things to make it.	[D2] = yes		
63	Ball	And the two things that you put together to make it were odd, right? Three and three are each odd?		[1.5.3] Addressing the issue and seeking closure	
64	Sean	Uh huh, and the other, the twos were even.			
65	Ball	[A] So you're kind of – I think Nathan said then that he wasn't talking about every even number, right, Nathan? [B] Were you saying that? [C] Some of the even numbers, like six, are made up of two odds, like you just suggested.			
66	Nathan	Uh-uh (agreeing with the teacher).	[D5] = yes		
67	Ball	Other people's comments?	[1.6] 6th cycle of routine	[1.6.1] initiation of 6th cycle (Action A2)	

APPENDIX 2A (*continued*)

Beliefs: Relevant Knowledge   Decision-Making

See Appendix C for the Belief Legend.
Beliefs L1-L6 are about learning
Beliefs T1-T4 are about teaching
Beliefs S1-S4 are about students
Beliefs C1-C5 are about classroom environments
Beliefs M1-M3 are about mathematics

Betsy begins interacting
with Nathan, at which point
Ball takes a step back to see
how the interaction will go.

As described in the narrative,
productive and respectful
student exchanges are a
major goal for Ball's classroom.
Beliefs related to such
exchanges are triggered, and
a "wait and watch" monitoring
routine is activated.

With Nathan's understanding
clarified, Ball has met the
objectives she set out to
achieve by taking the
"detour" in line 26. [D5] = yes,
so she returns to the
debriefing plan.

[D1] = yes and [D2] = yes.

As the narrative indicates,
Ball thought what she said
in line 65 clarified issues
for Sean. This completed
another cycle of the
debriefing routine.
[D5] = yes, so Ball asks
for more comments.

APPENDIX 2A (*continued*)

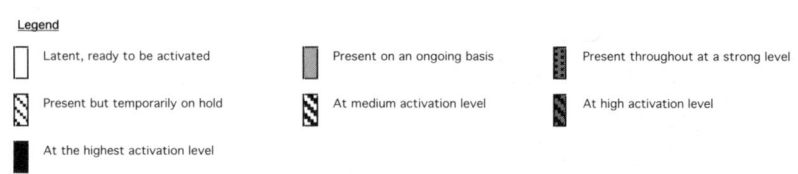

Legend

☐ Latent, ready to be activated ▩ Present on an ongoing basis ▦ Present throughout at a strong level

▨ Present but temporarily on hold ▨ At medium activation level ▨ At high activation level

■ At the highest activation level

APPENDIX 2B

High Priority Goals During This Lesson Segment

Overarching Goals

A. Reinforce community norms of respectful and substantive interaction, employing classroom-management techniques where appropriate to do so.

B. Pursue issues of mathematical substance, specifically those related to promoting mathematical discourse (including the collective consideration of mathematical claims or arguments) as they arise during classroom discussions.

C. Explore students' understanding of a topic or idea when their state of understanding is unclear to Ball.

Major Content and Learning Goals for This Lesson Segment

D. Debrief students on the previous day's meeting (and let the discussion take its course, as long as this course is consistent with other high-priority goals).

E. Focus on how students understand their own mathematical understanding and learning, in the context of (and using as data) the previous day's meeting.

Local Goals (in the Context of Specific Student Actions)

F. Have a student clarify, elaborate, or extend what he or she has said.

G. Highlight a student's statement (perhaps transforming it or adapting it) in service of top-level instructional goals.

H. Monitor student exchanges attentively (a) to determine whether intervention is necessary and (b) to reflect on them later, when appropriate.

I. Bring a student into the conversation by tying the conversation to something he or she has said previously.

J. Build a platform for work that the teacher intends to pursue later in the lesson.

APPENDIX 2C

General Description of Deborah Ball's Beliefs Active or Potentially Active at Start of Lesson on January 19, 1990

Beliefs About Learning

Mathematics learning includes the following:

(L1) developing a deep and connected understanding of mathematical content;

(L2) developing a particular set of mathematical values, including the idea that mathematics—

> should make sense,
>
> can be worked through logically;

(L3) learning to learn, including —

> reflecting on what one knows,
>
> reflecting on how one learns.

Mathematics learning occurs through the following:

(L4) participating in mathematical discourse, which includes—

> generating mathematical ideas,
>
> listening to others' mathematical ideas, and
>
> critically examining mathematical ideas, as individuals and collectively;

(L5) participating in a community that supports mathematical sense-making; and

(L6) articulating one's thoughts, orally and in writing.

Beliefs About Teaching

Teaching includes the following:

(T1) establishing, building, and maintaining a classroom community that enfranchises students and supports sense-making;

(T2) probing and making sense of students' understandings and beliefs;

(T3) giving highest priority to pedagogical strategies that—

> focus on student understanding and learning,
>
> focus student attention on content, reasoning, learning and community,
>
> provide students opportunities to express their ideas,
>
> have resolutions to mathematical issues emerge from rational classroom discourse rather than by virtue of teacher authority;

(T4) valuing students'—

> initiative,
>
> understandings, and
>
> ideas.

Beliefs About Students

Students—

(S1) should all feel enfranchised in the classroom community, free to express themselves;

(S2) have important ideas;

(S3) should be nurtured as individuals; and

(S4) can/should all behave as members of the classroom community.

Beliefs About Classroom Environments

A productive classroom environment—

(C1) enfranchises all students,

(C2) has norms of paying attention, including listening carefully to other students as well as to the teacher,

(C3) focuses on rich mathematical ideas,

(C4) reflects the sense-making values of the larger mathematical community, and

(C5) provides opportunities as described in "beliefs about learning" and "beliefs about teaching."

Beliefs About Mathematics

Mathematics—

(M1) is a sense-making activity;

(M2) includes certain content and process understandings,

e.g., "problem solving, communication, connections, and reasoning" and NCTM Content Standards;

(M3) involves a particular type of reasoning, including—

justifying claims on the basis of evidence and articulated assumption (definition), and

making clear and logical arguments.

Chapter 3

Accountable Argumentation as a Participation Structure to Support Learning Through Disagreement

Ilana Seidel Horn
University of Washington

The first two chapters in this volume focused, in different ways, on the teacher and the teacher's work—the myriad things a teacher does to ensure that a classroom functions well. In this chapter I engage in a complementary analysis, with a focus on a particular discourse structure that enables some of the classroom work to be done.

Why focus on classroom discourse? Classroom discourse has become an important site of inquiry and change in mathematics education (Lampert & Blunk, 1998; NCTM, 2000; Kieran, Forman, & Sfard, 2003). As students endeavor to make sense of mathematics, their talk and interaction in the classroom provide an important resource for the construction of meaning (Cazden, 2001; O'Connor, 1998). Cazden (2001) describes classroom discourse as comprising a crucial part of the "social plane" in which higher order thinking first appears before it becomes appropriated by individuals.

Argumentation has an especially important role in mathematical thinking (Lakatos, 1976; Lampert, 1990). If students are to engage in authentic mathematical activity while making sense of the subject, then argumentation and disagreements should find a way into the discursive practices of mathematics classrooms. In discussion-intensive classrooms, students can be, in a sense, apprenticed to argumentation and reasoning practices through their participation in the classroom (Lave & Wenger, 1991).

However, as teachers of discussion-intensive classrooms relate (Lampert, Rittenhouse, & Crumbaugh, 1996), such disagreements come with the social risk of losing face (Renkema, 1993) in front of peers. A question thus arises: How can classroom discourse be organized to support mathematical disagreements that (a) are intellectually productive, and (b) minimize social discomfort? In everyday interaction, disagreements are viewed negatively. Understanding how teachers support productive intellectual disagreements in classroom conversations is thus worthy of examination.

The data provided for this volume are ideal for investigating the foregoing questions. During the class session of January 19, 1990, Ball's students generate a "new" class of numbers while arguing vigorously about the nature of odd and even numbers. Ball manages to support the children's engagement in this activity while maintaining the social equilibrium of the classroom.[1] The class session is thus an exemplary instance of mathematical argumentation that preserves the intellectual community of the classroom.

This chapter focuses on describing the interactional organization of *accountable argumentation,* an activity embedded in whole-class discussions in Ball's classroom. An analysis of the classroom's interactional organization can foster insight into how this particular discussion managed to be intellectually productive without disrupting the equilibrium of the classroom.

Organization of Chapter

This chapter is organized in the following manner. First, I locate this analysis theoretically, tracing its connections to analyses of both classroom-participation structures and the discourse of disagreements. In the second part of the chapter, I explicate the classroom's interactional organization by describing a participation structure, which I call accountable argumentation, that supports mathematical learning through disagreement. I then analyze two disagreement episodes to illustrate accountable argumentation in action. Finally, I conclude with a discussion of accountable argumentation as a pedagogical and analytical resource for investigating mathematical teaching and learning.

THEORETICAL PERSPECTIVES

Participation Structures

A helpful approach in understanding how the class co-constructed the odd-and-even numbers during the session is to examine the classroom discourse practices. Classroom discourse practices are socially developed, patterned ways of using language, gestures, and representations, coordinated with understandings about the subject matter (Greeno, Benke, Engle, Lachapelle, & Wiebe, 1998). One analytic construct used to understand those patterned interactions is that of classroom *participation structures* (Philips, 1983/1993). Participation structures are interactionally emergent, providing and organizing resources for learning (Hanks, 1991). Participation structures allocate student involvement in classroom activities and produce a level of interactional organization within which the structuring of any single encounter is accomplished (Philips, 1983/1993, p. 79; see also Herrenkohl & Guerra, 1998). The structure, however, is highly contingent; its rules provide resources for participants to play the game according to their own strategies and are not simply obeyed (Lemke, 1990 p. 9). Some

---

[1] Please see Posner (2007/this volume, pp. 127–172) for another interpretation of the social viability of this dispute. Her analysis raises important concerns about the possibility of problematic social processes that may be enacted during this dispute.

examples of participation structures commonly found in classrooms are *whole-class discussion, group work,* and *teacher tutoring.*

On a descriptive level, those participation structures provide a way of seeing a given activity. On an analytic level, participation structures allow for the investigation of the relations between social actors and their interactions. That is, the examination of classroom participation structures supports the analysis of the relations among students, the teacher, and the various classroom activities, lending insights into the organizational possibilities and limitations for participation and learning. In studying participation structures, researchers seek to codify (a) those relations, (b) the corresponding positions, and (c) the normative expectations for appropriate conduct. From there, patterns in interaction among participants can be highlighted and analyzed (Goffman, 1981; Hanks, 1996). In educational research, analyses of the organization of classroom participation—and the way various classroom participants draw on, resist, and transform that organization—allow for context-specific descriptions of learning.

The Discourse of Disagreements

To understand disagreements in this classroom, I draw on three sources. The first source is Magdalene Lampert, who, along with her colleagues, has written about the role of disagreements in her discussion-intensive mathematics classroom (Lampert, 1990; Lampert, Rittenhouse, & Crumbaugh, 1996). Lampert and colleagues' analysis describes the social and personal tensions that arise as students engage in academic disputes, which the authors see as having a place in mathematical learning. They argue that engaging in disagreement is closer to authentic mathematical and scientific practice than the types of discourse that typify traditional classrooms. As the authors say, "by posing interpretable problems and encouraging disagreement, the teacher sets the stage for students to clarify their thinking and relate thought to communication" (p. 738). How that stage is set, the roles available to the players, and the discursive details of the drama of disagreements are not a focal part of their analysis but are explored in this chapter. Interestingly, in those discursive details lie a multitude of interactional strategies that participants employ to manage the social and personal tensions as they engage in disagreements.

Including disagreements in the teaching of mathematics is, in part, an attempt to bring mathematical pedagogy closer to the rigors of formal academic practice (Chazan & Ball, 1995; Lampert et al., 1996). Accounts of authentic academic disputation practices thus provide a useful framework for sorting out some of the complexities that arise during academic disagreements. Bruno Latour's (1987) work on this aspect of scientific practice provides a way to delve more deeply into those complexities. Specifically, his analysis provides a language for the strategies scientists use (e.g., *recruiting allies, reifying ideas through inscriptions*) and the positions they take (e.g., *dissenter*) in the course of disagreements. It turns out that the members of Ball's classroom, in their appropriation of academic argumentation, employ some of the same strategies and position themselves in ways similar to those used by professional scientists.

Finally, because the students are children who bring in their own personal understandings of how disagreements are managed, Marjorie Goodwin's (1991) linguistic ethnography of urban black children's talk sheds insight into the discourse of peer disputes. Although the students in Ball's classroom come from diverse ethnic backgrounds (see Appendix 2 of this volume), the analytic categories that emerged from Goodwin's fine-grained ethnographic work illuminate many of the relational stakes involved in children's arguments, as well as a detailed analysis of the strategies they employ in those interactions without the mediation of adults.

Bringing those three perspectives together, this chapter elaborates the role of language as a resource for positioning students in the course of mathematical disagreements in the classroom. Specifically, this analysis illustrates the participation structure of accountable argumentation.

DATA AND METHODS

Data

Deborah Ball provided access to her rich set of data from this classroom, which was collected as part of the Mathematics and Teaching Through Hypermedia (M.A.T.H.) Project. The primary data for this analysis were the videotape of the January 19, 1990, class session and a transcript of the session. In addition, to challenge and strengthen the analysis that emerged from the primary data, I also reviewed a transcript of the previous day's session; Ball's journals from the days leading up to, including, and following January 19; copies of the students' notebooks from January 19; and prior analyses of the day's events (Ball, 1998; Ball & Bass, 2000). Those secondary data sources were used to seek confirming and disconfirming evidence of the analysis presented here.

Methods

Methodological approach. Methods from both conversation analysis and sociolinguistics were employed in this analysis. Conversation analysis seeks to describe the underlying organization of social interaction. To do so, it requires an integrated analysis of action and local context, with the assumption that speakers constantly influence and constrain the conduct of their coparticipants. Because coparticipants in interaction rely on spoken utterances or other actions to interpret a situation, this, too, is the point of departure for conversation analysts (C. Goodwin & Heritage, 1990; Sacks, Schegloff, & Jefferson, 1974). The utterances and actions of the students and teacher comprised the primary data on which this analysis relies.

Sociolinguistics adds depth to the conversation about the analytic notion of context by using ethnographic methods to interpret interactions beyond the immediate and local context created by talk and action alone (M. Goodwin, 1991; O'Connor & Michaels, 1996). Thus, the analysis in this chapter seeks to interpret the spoken utterances and gestures of the class participants in the broader contexts of interaction, such as the previous day's class and the teacher's perspective as represented by her reflections on the class discussion. Although the interpretations of the context did not come from traditional ethnographic data, the rich and various data that Ball's group provided deepened the understandings of the multiple contexts that influenced the local interactions.

Data analysis. In general, data analysis proceeded in a recursive and nonlinear fashion. For narrative simplicity, I describe three phases of analysis, despite the fact that these phases blurred in time. This description of phases should be taken as an outline of my analytic logic rather than a strict chronology of the analytic process.

In the first phase of the analysis, the video was repeatedly reviewed alongside the transcript provided by Deborah Ball (Appendix 1 to this volume). The disagreement episodes were coded in the original transcript, and are presented in Table 3.1. While reviewing the disagreement episodes across the class session, I took note of the disagreement episodes in which students managed their exchanges with one another with little or no teacher mediation. I considered those interactions as examples of *socially viable* disagreements among peers. I wanted to further understand how the students managed those interactions, especially since everyday disagreements are often socially uncomfortable and are thus frequently avoided. In addition, during the repeated viewing phase, I annotated the transcript with notes about participants' location, intonation, and gestures that amplified or shifted their meanings in ways not represented by their words alone.

As a part of that work, I set about categorizing the disagreement episodes. To decide whether a disagreement was *sustained,* I looked not so much for the number of turns within it but rather whether the argument maintained a *mathematical focus.* As previously mentioned, I also looked at the teacher's role (did the teacher strongly scaffold the disagreement, or did the students manage the disagreement with little teacher intervention?). I also took note of which students initiated the disagreement, as well as to whom the disagreement was directed.

In Phase 2, I selected two episodes for close analysis. Because I was interested in the *social viability* of those student-student disagreements, I focused on episodes that were not strongly scaffolded by the teacher. To help me understand differences in how the disagreements were managed by students, the two episodes differed in how successfully the argumentation was sustained. For this phase, I wanted to pay close attention to language and action, so I used a thoroughly revised transcript of the episodes that included more of the false starts, repeated words, and acknowledgment tokens ("mm-hm"'s), using transcript conventions common to this type of analysis (Ochs, 1979; see Table 3.2). Those revised transcripts are excerpted in this chapter. The additional utterances, along with notes about gesture, physical location, and intonation, became important clues for understanding and interpreting interaction in context. The trade-off, of course, is that the transcripts excerpted in this chapter may be less readable, but for those accustomed to the notation, they provide greater access to the sound and pace of the conversations.

Table 3.1.

Accountable Argumentation Episodes* Found in the January 19, 1990, Class Session

Episode	Transcript turn numbers	No. of turns	Topic of controversy	Sustained?	Strong teacher scaffolding?	Initiator	Principal
1	10–19	10	Is zero even?	No	No	Sean	Sheena
2	45–58	14	Do all even numbers make even numbers?	Yes	No	Betsy	Nathan
3	69–78	10	Is six an odd number?	Yes	No	Cassandra	Sean
4	131–143	13	Is six an odd number?	Yes	Yes	Tembe/Students	Sean
5	149–165	14	Is six an odd and even number?	Yes	No	Mei	Sean
6	191–193	3	Is six an even number?	Yes	Yes	Sean	Ofala
7	298–309	12	Can six be even and odd?	Yes	Yes	Cassandra	Sean
8	367–370	4	Can ten be odd?	Yes	Yes	Ofala	Sean
9	390–426	37	Which numbers can be odd-and-even numbers?	Yes	Yes	Riba	Sean
10	431–445	15	Does an odd number of groups of two make an odd number?	No	Yes	Ofala	Sean

*The two highlighted episodes are closely analyzed in this chapter.

Table 3.2.
A Summary of Transcript Conventions, Adapted From Ochs, 1979

/	Self-interruption
=	No gap between utterances
(.), (5)	Very slight pause, five second pause
//,]	Beginning of overlapping utterances, end of overlapping utterances
,	Low rise in intonation
?	High rise in intonation
::	Marks lengthened syllable, each : equals one "beat"
.	Low fall in intonation
italics	Marks stress
capital letters	Increased volume
-h, h, (h)	In-breath, out-breath, laughter
(??), (cow)	Unclear reading, tentative reading
(( ))	Marks other voice qualities or actions

The revised transcripts of the focal episodes were analyzed inductively (Glaser & Strauss, 1967) to seek out a participant structure that organized students' disagreements. As was described in the "Participation Structures" section, participant structures codify relations among participants through an analysis of positions taken and normative expectations for conduct. To understand the organization of those disagreements, codes were developed for the elements of a participant structure: *interactional roles,* as well as *norms and expectations. Roles* described the interactional work participants did while engaging in a disagreement. *Norms* referred to the standards or patterns of social behavior that were accepted in, or expected of, the class but were more tacit than the explicitly stated *expectations.* Those codes were revised and revisited throughout the coding process. Once the revised codes satisfactorily described the interaction in the two focal episodes, they were then applied to the other disagreement episodes and Ball's journals. When the categories captured the range of disagreements, they were deemed satisfactory.

In the final phase, I developed a framework for how roles, norms, and expectations functioned in interaction to sustain (or not) students' disagreements. Looking at argumentation in interaction, I developed the following two insights: (1) the delegation of interactional roles had predictable patterns; and (2) the use of *prior conversations* was crucial to the way disagreements unfolded. After integrating those insights into the analysis, I applied the framework and insights to all the disagreement episodes and found no inconsistencies in the way the framework and insights captured them.

In the next section, I explicate the structure of accountable argumentation. Then I illustrate how it works in action through a close analysis of the two closely transcribed episodes of student-student disagreements. The first episode illustrates an instance in which accountable argumentation is not sustained, and the second episode shows an instance in which accountable argumentation is sustained and generates new mathematics for the class. Ball's minimal scaffolding in these episodes presumably increased the social stakes for the students, making these episodes salient for my particular questions.

FINDINGS: ACCOUNTABLE ARGUMENTATION IN ACTION

Overview of Findings

The following analysis provides an abstract and then an enacted description of accountable argumentation. First, I describe the abstract structure of accountable argumentation by delineating the norms, expectations, roles, and uses of history that define it. Next, I illustrate accountable argumentation by presenting the analysis of the two focal disagreement episodes. Episode 1 occurred during the first six minutes of the January 19, 1990, class session and illustrates how accountable argumentation, compared with other participation structures in the classroom, supports mathematical reasoning and learning. Episode 2, a disagreement that occurs later in the class session, shows how the interactional roles in accountable argumentation constitute a resource for the development of mathematical ideas. During Episode 2, the odd-and-even concept is made more general and, therefore, more mathematical.

Accountable Argumentation: The Abstract Structure of Participation

Accountable argumentation is a participation structure, embedded in whole-class discussion, that organizes the public disagreements among students and provides interactional resources for clear mathematical reasoning and the production of mathematical generalizations. I call this structure *accountable* argumentation because it behooves participants to be responsible for many elements of the discussion. Participants are held accountable for following their peers' thinking, remembering previously developed mathematical ideas, and producing rigorous reasoning and justification.[2]

Norms and expectations. Although I coded norms and expectations separately in the data analysis, I report them together here for reasons of space and simplicity. In the analysis, I coded regularities in classroom interaction as *norms* (Yackel & Cobb, 1996; McClain & Cobb, 2001), whereas explicit statements that sought to structure participation were coded as *expectations*. This distinction is not consequential for understanding the participation structure of accountable argumentation, as what may look like norms in one class session may be explicitly stated as expectations in another.

---

[2] For a similar use of the word *accountable,* please see Resnick (1999).

Norms and expectations contribute to the organization of participants' activities by both constraining and providing a resource for those activities. Since accountable argumentation is embedded in the whole-class discussions in Ball's classroom, the norms and expectations described are common to both participation structures.

Listed below are some norms and expectations that structurally distinguish accountable argumentation from what might be thought of as more typical whole-class discussion.[3] Each one is briefly described or illustrated using excerpts from the class session. Participants in this classroom are expected to do the following:

1. *Attend to contributions in a whole-class discussion*

    Ball: And could you listen to one another's comments, so that we can, um, (1) benefit from what other people say?

    ★★★★★★★★★★★★★★

    Ball: Tembe, are you listening to Riba? She's trying something else now.

2. *Value disagreements and understand that they may not (and need not) be resolved*

    Ball: I understand that this is a problem and that we didn't … settle it. We're probably not going to settle it.

    ★★★★★★★★★★★★★★

    Ball: Anybody want to comment on this, or is this one of the examples of something we've got to let rest a little bit now?

3. *Support thinking during discussions through the allocation of attention and time.*

    Ball: Riba, can you watch what Mei's doing?

    Mei: Let's say that I have (*7 second pause*). Let's see.

4. *Have a justified position in a discussion*

    Ball: What are you convinced by?

    ★★★★★★★★★★★★★★

    Sheena: I don't, I don't agree with that.

    Ball: Why?

5. *Act on or defend a position in a discussion*

    Sean: I disagree because, um, because what= what two things can you put together to make it?

    ★★★★★★★★★★★★★★

    Ball: You're going to have to help because she's talking to you and I'm not sure what she's trying to show.

6. *Use terms from the mathematical and academic registers.*

    Tembe: Prove it to us!

---

[3] I am contrasting accountable argumentation with two forms of whole-class discussion, the first being typical whole-class discussion found in U.S. classrooms (Stigler & Hiebert, 1999) and the second being whole-class discussion in Ball's class. Accountable argumentation is embedded in the latter and thus shares some characteristics with whole-class discussion in that setting.

****

Ofala: My conjecture, I think it's always true.

7. *Respectfully respond to others' positions in a discussion*
    Sheena: Could you repeat what you said, please?
    ****

Sean: I didn't think of that that way. Thank you for bringing it up.

8. *Revise a position in light of new questions or convincing evidence.*
    Ball: [...] Sheena commented that (.) it was good to have the two classes together because she heard an idea that she hadn't thought about and it made her think about and even revise her own idea when she was (.) in the meeting yesterday.

****

Nathan: First I said zero was even but then I guess I revised so that zero, I think zero is special.

These norms and expectations are communicated consistently through the participants' interactions. Often, violations are noted explicitly, as when a student is not listening attentively (e.g., "Were you not listening to this just now?") or when somebody has not adequately substantiated a controversial position (e.g., "Prove it to us!"). Note that the norms and expectations of accountable argumentation vary radically from those of traditional mathematics classrooms, in which the authority of the text or teacher supersedes students' valuation of their own thinking (Schoenfeld, 1988).

Ball, in her role as teacher, takes primary interactional responsibility for the first four norms and expectations during the class session examined for this analysis. Ball and the students themselves hold one another accountable for the last four. Perhaps over time, all participants learn to hold themselves and one another increasingly accountable for meeting all these norms and expectations.

Taken together, these norms and expectations support the social viability and mathematical productivity of disagreements in Ball's classroom. Norms and Expectations 4 through 6 shape the rigor of the disagreements. By expecting students to not only take a position in an argument but also defend that position using academic language, the participation structure creates a high-press environment for student learning (Kazemi & Stipek, 2001). Norms and Expectations 6 through 8 help support the social viability of classroom disagreements. The academic language used for these disagreements may help distinguish them from everyday playground disputes, while the polite tones and normalization of "revising" allow students to maintain a civil tone and save face.

Roles. The roles described subsequently are those that are taken up and enacted during accountable argumentation sequences during whole-class discussions. This description of roles reflects an inductive analysis of the discursive practices in the classroom, and their labels have been borrowed from other accounts of disagreements in both discourse analysis (Goffman, 1974) and the social studies of science (Latour, 1987).

The following roles are available to participants engaged in accountable argumentation:

1. *Principal of a controversy* (Goffman, 1974, p. 517): A person held accountable for a position that others disagree with or question.

2. *Dissenter* (Latour, 1987): A person who takes an opposing position to the principal's position.

3. *Ally* (Latour, 1987): A person (or thing) who (or that) supports the principal's position.

During disagreements, participants may take up any of these first three roles. Oftentimes, they blend them with one or more of the following:

4. *Questioner:* A person who asks questions, especially of the principal, perhaps because of confusion or an uncertain stance.

5. *Reasoner:* A person who provides an exposition of reasoning.

6. *Listener:* A person who listens to the arguments.

7. *Norm-maintainer:* A person who explicitly evokes norms during conversation.

8. *Clarifier:* A person who clarifies or summarizes another participant's statement. As will be illustrated in Episode 2, clarifying may be the first half of a revoicing move (O'Connor & Michaels, 1996).

In this description, I do not wish to suggest that these roles are explicitly delegated or stable over time. The roles are occupied through highly contingent, negotiated social processes that will be explicated in the following section.

Negotiating roles in interaction. Participants in discussion can utilize these roles as resources in several ways. First, they can *assume* (Goffman, 1974) the various roles for themselves through their talk. For example, by beginning a speaking turn with the statement "I disagree with Joe," a participant assumes the role of *dissenter*. Second, participants can *design* (Sacks, Schegloff, & Jefferson, 1974) others in these roles. With the same utterance ("I disagree with Joe"), the speaker designs Joe as a *principal of a controversy.* Once that role has been *ratified* (C. Goodwin & Heritage, 1990) by the designated principal—Joe might silently turn and face his interlocutor or, more explicitly, say, "Okay, what's your question?"—the stakes are then raised for Joe to act on or defend the controversial position for which he is now accountable. Finally, participants can *animate* (M. Goodwin, 1991) others in these roles. In the midst of summarizing an argument, for instance, a speaker may juxtapose the positions of two participants to point out their agreement on an issue. In this situation, the two participants are animated as *allies.*

History

Through animation, speakers can strategically recycle past utterances or positions in the present interaction. By animating a past speaker's position, current speakers can recruit allies or take positions of dissent. Additionally, because participants are frequently animated as holding positions from previous class sessions, this continuity intensifies participants' day-to-day accountability for their positions on a topic. Participants may be *yoked* (O'Connor & Michaels, 1996) into a discussion when their past positions are recycled, connecting them with present discussions through the animation of a current speaker.

Some aspects of history are also brought into the present through the use of inscriptions (Latour, 1987), which are representations that reify ideas, serve as pedagogical devices, support talk over and about them, and support the recycling of past arguments (Greeno & Hall, 1997; Roth & McGinn, 1998). Inscriptions provide important mediating resources for argumentation. In fact, inscriptions themselves are often animated as allies during reasoning sequences, as is illustrated in Episode 1.

In this classroom, a few inscriptions are featured in the course of the discussion. These instances are representations of numbers from which evenness or oddness may be derived, including the number-line representation, a hash mark representation (for example, 4 would be represented as ||||), and a "cookie" representation of numbers (O O O O). These representations support different understandings of even and odd. That is, the number line highlights the alternation of even and odd integers, and, although both the hash mark and cookie representations allow for "grouping by twos," individual cookies are more easily "split in half." These inscriptions and their entailments are summarized in Table 3.3.

Table 3.3.
Inscriptions* Used During Whole-Class Discussion on January 19, 1990

Name	Representation	Highlights
Number line	0 1 2 3 4 5 6 7 8	Alternating pattern of even and odd
"Cookies"	O O O O O	Groups or pairs; splitting in half
Hash marks	| | | | | | | | |	Groups or pairs

*Each representation highlights a different feature of even and odd numbers.

Accountable Argumentation Illustrated

In this section, accountable argumentation is illustrated through the analysis of two episodes of interaction in the context of whole-class discussion. The features of the abstract structure—that is, the norms and expectations, roles, and uses of history—provide the conceptual language for this analysis. Episode 1 shows an instance in which accountable argumentation is neither sustained nor productive. In the analysis of this episode, I contrast accountable argumentation with other participant structures organizing the students' disagreement to illustrate the opportunities for learning that accountable argumentation provides. Episode 2 shows an instance in which accountable argumentation is both sustained and productive.

In the analysis of this episode, I show how the roles in accountable argumentation support students' mathematical thinking.

Episode 1: A Disagreement About Zero

Overview of participation structures in Episode 1. This episode traces a disagreement about zero as enacted by two students in Ball's class, Sean and Sheena. The episode starts in a whole-class discussion format in Part 1, with Sheena discussing with Ball her current thinking about the evenness or oddness of zero after the previous day's meeting. An accountable argumentation sequence begins in Part 2, when Sean disagrees with Sheena's position, and continues in Part 3, when Sheena defends her position. Accountable argumentation gives way to a peer-dispute format (M. Goodwin, 1991) at the end of the episode, at which point Ball steps back in and redirects the activity to a whole-class discussion. In addition to illustrating some important features of accountable argumentation, this episode shows the contiguous relationships of the participation structures in the classroom and the fluid transitions between them.

Part 1: Reflecting on a position. At the beginning of the class session, Ball opens up the discussion by asking that students reflect on the previous day's meeting with the fourth graders on the topic of zero's oddness or evenness. Structurally, this format is a whole-class discussion with the teacher allocating turns of talk and designating the topic, which is "reflection on yesterday's meeting." Sheena, in dialogue with Ball, goes public with the position she has arrived at:

5   *Ball:*   Was there an example of something yesterday that (.) you understood a little bit more (.) during the //meeting?]

6   *Sheena:*   //Well:], I didn't think that zero was (.) zero, um even or (.) odd until yesterday they said that it could be even because of the ones on each side is odd, so that couldn't be //odd.]

7   *Ball:*   //Hmm.] So y=

8   *Sheena:*   =So that helped me understand it.

In this sequence of dialogue, Ball supports Sheena in taking a specific position on the topic of yesterday's discussion by asking her to provide a specific example of something she "understood a little bit more." In Turn 6, Sheena recycles an argument linked with the number-line inscription to substantiate her reasoning ("because the 1's on each side is odd, so that couldn't be odd"), bringing part of yesterday's discussion into the present interaction. She describes being convinced by something "they said" (Turn 6) that "helped [her] understand it" (Turn 8), conforming to the expectation that her position be reasoned.

Part 2: Initiating accountable argumentation. When Sean is called on as the next student to speak, he reorganizes the interaction from a discussion reflecting on yesterday's meeting to accountable argumentation:

11   *Ball:*   Other people's comments? Sean?

12	*Sean:*	Um, I, I, I just want to say something to Sheena =when sh/ what she said about um (.) that (.) that one, um zero has to be an odd/an even number bec/ I disagree because, um, because what= what two things can you put together to make it?
13	*Sheena:*	(1) Could you repeat what you said, please?
14	*Ball:*	((*speaks to Betsy and asks her to listen*))
15	*Sean:*	(1) Okay, um, I disagree with you because (.) um, if it was an even number, how/what two things could make it?

Perhaps because he is redirecting the activity from reflection to accountable argumentation, Sean somewhat haltingly prefaces his disagreement in Turn 12. In this preface, Sean addresses Sheena indirectly, using her name ("I, I, I just want to say something to Sheena"). After Sheena directly addresses him (Turn 13) and he takes a slight pause, Sean addresses her in the second person ("Okay, um, I disagree with you"). This pattern of assuming a position of dissent by starting in third person ("Sheena") and changing to second person ("you") only after the dissent has been ratified by other participants occurs elsewhere during the class session. This occurrence may mark a transition point from addressing the teacher in whole-class discussion to addressing a peer in accountable argumentation. Possibly this strategy also helps students manage the relational discomfort of disagreeing with one another.

To understand this point more fully, consider the range of possible alternatives. On the one hand, if a dissenter initiated a dispute in the second person ("you"), the dispute might immediately seem personal. On the other hand, if a dissenter were to address a vague third person ("somebody said"), an expansive, generalized principal ("people who think…"), or simply the idea in general ("about the idea that zero is …"), the controversy's trajectory is more limited because no one is the obvious respondent to the dissenting point. By comparison, the initiation of a dispute in the third person avoids the harsh confrontational tone of the second person while sustaining the accountability of the principal to any position he or she has publicly taken.

After Sean designs Sheena as the principal of a disagreement, Sheena ratifies the role of principal and now takes on the rights and obligations of that role. Principals in this classroom, for example, are obligated to defend their position in a disagreement by explicating their reasoning. They also have the right to assume a central position in the classroom, both physically and discursively, as will be illustrated in this example (and in Figure 3.1).

In Turn 15, Sean formulates a question for Sheena. He merges the roles of dissenter and questioner, employing a common way for dissenters to account for their own positions in this classroom. Sean supports his dissent by asking a question that reflects his understanding of even numbers. That is, his statement reflects an understanding that even numbers are made up of two like numbers, an understanding supported by the circle or hash mark representations. By posing a question, Sean also designs Sheena as having a reasoned position from which she can respond and underscores her obligation to defend her now-controversial position.

When students have stepped into the roles of principal and dissenter, Ball no longer manages speaking turns from her position of teacher. The student-student dialogues during whole-class discussions occur during accountable argumentation sequences, once the roles of principal and dissenter have been assumed and ratified in interaction. This aspect of accountable argumentation decentralizes the teacher's authority to mandate turns, allowing students to manage some of their own interaction.

In this sequence of dialogue, Ball quietly works as a norm-maintainer by asking Betsy to listen to the discussion. In doing so, she designs Betsy as a listener in the discussion and underscores the expectation of attending to the arguments in a controversy, even when one does not hold the floor.

Part 3: Defending a position. Sheena then goes on to defend her position:

16 *Sheena:* Well, I could show you it. (.)

((*Moves toward the chalkboard and points to the number line with a yard stick*))

Um, I forgot wh/what his name was, but yesterday he said that this one

((*Points to the 1 on the number line*))

and each, this one is odd and this one

((*Points to the −1 on the number line*))

is odd, so this one

((*Points to the zero*))

has to be even.

Figure 3.1. Sheena defends her position at the number line.
(Source: videotape of Deborah Ball's class, January 19, 1990, M.A.T.H. Project.)

Sheena, in the position of principal, must justify her position on zero. To support her effort, she moves into what is typically the teacher's position at the front of the room near the chalkboard, a location of interactional authority and visual centrality.

Additionally, she recruits two important allies from yesterday's meeting, appealing to the history of this controversy over zero. During the exposition of her position, Sheena first allies herself with one of the fourth-grade boys ("I forgot wh/what his name was"), giving him authorship of her reasoning over the number line ("but yesterday he said"). This move accomplishes two things for Sheena. First, the alliance with an older (and perhaps therefore more authoritative) boy shows that Sheena's position is not anomalous but rather shared by another, perhaps more knowledgeable, peer. Second, this move alleviates some of the relational pressures of the principal position, inasmuch as the unnamed boy is the author of the argument and not Sheena herself.

Sheena also invokes the number-line inscription (and, as a consequence, the argument it represents) as an ally to substantiate her position, even though Sean's question actually refers to a representation of even numbers (Turn 15: "what two things could make it?") that relies on the notion of even numbers being comprised of same-sized pairs. They are thus engaged in cross-representational discussion, because the alternating pattern of the number line and the two-groups representations reflect different aspects of evenness.

Part 4: From accountable argumentation to peer dispute.[4] At this point, perhaps because of the seemingly incommensurable representations of evenness, the discussion shifts and takes on a format sometimes found in peer disputes (M. Goodwin, 1991). In the peer-dispute format that emerges, Sean and Sheena echo each other's utterances; their exchange is an example of format tying (M. Goodwin, 1991, p. 177), a reciprocal action that transforms Sheena's meaning and escalates rather than explicates the dispute.

17 *Sean:*   But that doesn't mean it always is even.

18 *Sheena:*   It *could* be even.

19 *Sean:*   It *could* be, but

20 *Sheena:*   I'm not saying that it *ha::s* to be even. //I meant that it could be.]

21 *Sean:*   // You said it was]

     ((*Sheena puts the yardstick back in the chalk tray and looks up at Ball.*))

22 *Ball:*   (2) Before we take this up again, I underst/I (.) I understand that this is still
            a problem = and that we didn't a/we didn't settle it. (.) ((*Sheena returns to
            her seat.*)) We're probably not going to settle it. (.) U:m (.) there's a lot of
            disagreement about this issue, right? And you saw that the fourth graders

---

[4] By drawing this contrast, I do not wish to suggest that peer dispute never has features of accountable argumentation. Goodwin (1990) describes, for example, the uses of proof and justification that are often employed within peer disputes. The polite tones, public forum, classroom setting, and appropriation of the mathematical register do seem to distinguish accountable argumentation from typical peer disputes.

who have been thinking about this for a long time also disagree about it, don't they? I'm still kind of interested (.) um:, in hearing some more comments about the meeting *itself*. [...]

Sean's challenge in Turn 17 recalls a category of numbers that emerged in the prior day's discussion. In conventional mathematics, even and odd are mutually exclusive categories of integers. During the previous day's meeting, some students held the position that zero is perhaps both odd and even, a special number that straddles categories. The "odd and even" category eliminates the mutual exclusivity of even numbers and odd numbers, which the number-line representation reifies.

In Turns 18 through 20, Sheena weakens her commitment to her position that zero is even by hedging ("It *could* be even"), further distancing herself from her original position in the face of Sean's dissent. Her position in Turn 6 (zero "couldn't be odd") has been transformed in the course of this disagreement to the statement in Turn 20. Sean's response (Turn 19), a partial repetition of Sheena's prior talk, does not represent further reasoning about the issue at hand but rather serves mainly to escalate the dispute. Sean's addition of "but" might index his previous Turn 17 ("But that doesn't mean it always is even"), thus continuing to push his position without providing further justification. At this phase of the dispute, neither Sheena nor Sean formulates explanations to accompany her or his assertions. This *exchange and return* sequence (M. Goodwin, 1991, p. x) indicates that the students are no longer focused on the validity or invalidity of each other's statements. The peer-dispute format that now organizes the exchange subsumes accountable argumentation's emphasis on reasoned positions.

Sheena eventually retreats from her original stance and replaces the yardstick in the chalk tray (Turn 20), signaling the termination of the dispute, even though no explicit resolution has been reached.[5] Sheena continues to stand at the center of the room, near the chalkboard in the physical position of principal, although she has discarded her primary tool. This action signifies an ambiguous moment of role transition, open for design by others. Sheena's glance at Ball invites the teacher's next move.

After a two-second pause, Ball steps in to normalize the unresolved ending (Turn 22). When she does so, Sheena picks up the cue to relinquish her role as principal and returns to her seat. Ball says aloud that "we didn't settle it," referring to the disagreement about zero. Her use of the pronoun *we* reestablishes the dispute as a collective issue for the whole class to grapple with, clearly demarcating it from a personal conflict between Sean and Sheena. Ball continues to underscore the unresolved status of this issue, emphasizing that "we're probably not going to settle it," before she redirects the activity to whole-class discussion. This move serves to normalize disagreements in her classroom while moving the students back to her original agenda.

---

[5] Such closings provide what Goffman (1971, p. 140, as cited in M. Goodwin, 1990) calls "ritual equilibrium."

Discussion of Episode 1. This disagreement illustrates some of the social stakes of publicly stating one's position on an issue. Sheena, in the context of reflecting on the previous day's meeting, is encouraged by Ball (Turns 5 through 8) to have a principled position about zero's oddness or evenness. In that context, she elaborates her stance to Ball to fulfill the expectation of supporting her position. In doing so, however, she makes her position public, which then opens her up to dissent from others. When Sean acts as a dissenter to her position, she manages the pressure of the role of principal of a controversy by recruiting allies and distancing herself from authorship of her position. The cross-representational discussion escalates the disagreement. At this point, the two students abandon accountable argumentation for a peer-dispute format that does not rely on reasoned positions, at which point Sheena hedges and eventually retreats from her stance.

When Ball steps in and redirects the activity to whole-class discussion, she appears to be choosing to return to the original topic and format that she set out (a whole-class discussion reflecting on yesterday's meeting). Had she wished to pursue the disagreement further, accountable argumentation could have continued. Although Sean and Sheena had arrived at an impasse, possibly a *clarifier* or a *questioner*—whether from Ball or another student—could have entered the conversation to point out, or ask about, the mismatch between their respective representations. Ball's choice to end the accountable argumentation sequence and leave the dispute unresolved is noteworthy. She prioritizes her other objectives for that day's lesson over the resolution of this disagreement.

Even though the disagreement ends without a visibly reasoned resolution, the structure of accountable argumentation still supports mathematical learning in this episode. Comparing, for example, Sheena's first statement in her position on zero (Part 1, Turn 6) with her explanation as principal (Part 3, Turn 16), we see that the latter explanation is more clearly stated. Standing at the center of the room, Sheena's communication has a different purpose: she is no longer communicating her reflections on yesterday's meeting to her teacher; instead she must defend herself in front of the whole class because her position has been called into question. From this central location, she has the number-line inscription and yardstick to use as resources to support her justification. In this example, we see how accountable argumentation supports Sheena's mathematical learning. First, by shifting the audience from a sympathetic teacher to a dissenting peer, accountable argumentation provided a compelling impetus for her to become more articulate and formal about her position. Additionally, by entitling Sheena, as principal, to the interactional center of the room—with its visible representational resources—accountable argumentation provided a means for her to step into a teaching position from which she can access and display inscriptions to support her justification.

Accountable argumentation as a participation structure uniquely supports this learning. Had the class' activity retained the organization of whole-class discussion, Sheena would not necessarily have needed to repeat her position or find a way to articulate it more clearly. Similarly, the linguistic and interactional resources of accountable argumentation supported the productive disagreement in Parts

2 and 3 of this episode, during which Sean states his disagreement and Sheena defends her position. In contrast, the peer-dispute format in Part 4, by shifting the focus away from evaluating the validity of one's argument to posturing and hedging, does not support displays of reasoning.

Episode 2: A Disagreement About Six

The controversy about six described in the introduction of this volume consumes much of the class session. It begins when Sean publicly states that he thinks six is an odd-and-even number, a statement that quickly incites dissent from his peers. In the process of arguing this controversy, the class participants coconstruct the odd and even numbers.

By tracing the progression of this idea, we can see how accountable argumentation provides the interactional resources for the development of the odd-and-evens. With Sean taking the role of principal in this controversy, the other participants in the class, through their dissent, collaboratively transform his utterance from a particular statement about six (that it could be an odd and an even number) to a general class of numbers. This movement from the particular to the general is quintessentially mathematical (Jurow, 2004) and, as is argued here, is a type of transformation supported by accountable argumentation.

Although many participants undoubtedly contributed to the discussion and, more specifically, to the construction of the odd-and-evens, I focus on one interaction between Sean and Mei, in which accountable argumentation supports the transformation of the statement about six.

Overview of roles in Episode 2. Whereas Episode 1 contrasted accountable argumentation with other participation structures in its capacity to support student learning, Episode 2 illustrates the ways in which the roles of accountable argumentation may help support generative mathematical thinking. In Part 1 of this episode, Sean, as a principal of the controversy about six, is struggling to articulate his position when Mei enters the interaction to help him clarify the nature of the category of odd-and-even. In Part 2, Mei shifts roles and acts as a dissenter, although she does so by appropriating Sean's perspective. Proceeding in this manner, she justifies her dissent in Part 3 by acting as a reasoner and generating a second example of odd-and-even. Her dissent from within Sean's viewpoint, it turns out, is mathematically generative, and the roles of accountable argumentation support this work.

Part 1: Clarifying. Sean, as the principal of this controversy about six, is engaged in an accountable argumentation sequence. He is standing at the board trying to prove his statement to his current dissenter, Tembe:

| 1 | Sean: | Because, because see this, there's two ((*drawing*)) number two over here, put that there. Put this here. |

OO|OO|OO

There's two, two, and two. And that would make six.

2   *Mei:*   *I* think I know what he's *say*ing.

3   *Tembe:*   Which is *even,* Sean.

4   *Ball:*   Mei?

5   *Sean:*   Yeah, and it could be (odd).

            ((*Sean walks away from the board*))

6   *Ball:*   Could you stay there? People have some questions for you.

7   *Mei:*   I think what he is saying is tha:::t.

            -h, it's almost, see/

            ((*Mei stands up in her seat*))

            I THINK what he's saying is that

            ((*Sits back down, grabbing back of seat*))

            you have THREE

            ((*holds up three fingers*))

            groups of TWO.

            ((*holds up two fingers*))

            And three is an ODD number -h

            ((*waving two fingers; rotates to face Ball*))

            so (.) SIX can be an odd number a::nd a even number.

Note that Ball maintains Sean at the center of the classroom (Turn 6). Ball tells him to stay at the board, the physical center of the classroom discussion, and explains that people have questions for him, highlighting his corresponding interactional centrality. The expectations of the principal role are maintained in her request: Sean must continue to stand up—quite literally—for his position. After his unsuccessful attempt to explain his position to Tembe, he seeks to abandon his physical (and perhaps some of his interactional) centrality.

In Turn 2, Mei steps in as a clarifier. Once her turn is ratified, she begins in Turn 7 what O'Connor and Michaels (1996) call a *revoicing* move. Revoicing is "a particular kind of reuttering (oral or written) of a student's contribution—by another participant in the discussion" (p. 71), which then opens up a slot in conversation for the original author of the statement to agree or disagree. In her role as clarifier, Mei does the first part of the revoicing move: she reformulates Sean's statement about six, highlighting his focus on the fact that three, the number of groups of two, is an odd number. In describing the pattern, Mei's reformulation not only clarifies but starts to generalize Sean's statement.

Ball then finishes the revoicing move by opening up the slot to Sean for confirmation or disconfirmation. O'Connor and Michaels explain that this part of the revoicing move "ultimately credits the contents of the reformulation to the student" (p. 71). Thus, Sean, whose talk has been animated and reformulated by Mei, maintains his participation status in the conversation as the *originator* of the statement; as such, he is positioned to confirm or disconfirm Mei's interpretation of his statement.

Part 2: Dissent. Mei then changes her interactional role. Ball asks her if she disagrees with Sean, designing her as a dissenter, a role that Mei readily assumes:

13 *Mei:*   Yeah, I disagree with that because -h

                ((*Stands up facing Ball*))

                it's not acco::rding to like, -h/

                ((*Pushes chair under table*))

                Here. Can I show it on the board?

14 *Ball:*   ((*pacing behind Mei's group's table*))

                Um hm

                ((*nods*)).

15 *Mei:*   ((*walking toward board, where Sean stands, leaning*))

                It's *not* according

                ((*arriving at board, picking up chalk*))

                //to like how many

                groups it is.]

                ((*pointing to Sean's diagram*))

Mei positions herself as a dissenter in relation to the reinterpreted statement about six. In Turn 13, Mei ratifies the position of dissenter by disagreeing "with that"—Sean's position, not Sean himself. This tactic is another strategy dissenters employ to manage the potentially personal feelings that come along with disagreement.

In assuming the role of dissenter, Mei immediately begins moving in to support her position. In Turn 13, she gets out of her chair while beginning to state her opposition. She seeks Ball's permission to go to the board, where Sean has been positioned in Turn 6. Moving and talking simultaneously, Mei states her point: "It's *not* according to like how many groups it is." By naming what she disagrees with, she further generalizes Sean's statement about six to describe a characteristic (the number of groups of two) that could also be found in other even numbers. Mei thus voices her dissent by arguing on more general grounds than Sean himself articulated.

Importantly, Mei's work as clarifier helped open a path for her dissent. Recall the previous disagreement between Sean and Sheena in Episode 1, in which Sean's dissent employed a different feature of even numbers than that on which Sheena's position was based (pairs of like numbers versus the alternating pattern on the number line). In contrast, Mei first establishes the grounds for Sean's thinking about six and then disagrees with him on his own terms, in what might be thought of as *empathic dissent.* That is, she projects herself into his perspective in an attempt to understand him better and continues to argue from within that perspective.

Part 3: Generalization through a counterexample

16 *Ball:*   //Riba, can you watch what Mei's doing?]

17 *Mei:*   Let's say:: that I ha::ve/ (7)

               Let's see.

               IF YOU CALL *six*

               ((*points to Sean's drawing of six*))

               an odd number, why don't =

               ((*facing Sean*))

18 *Sean:*   ((*under his breath*)) = Or it could be an even.

               ((*Standing at board, legs crossed*))

19 *Mei:*   ((*quietly, facing Sean*)) Let's see (if I find). (3)

               Let's say ten. One, two . . . ((*draws*))

               And here are ten circles. (1)

               ((*underlines circles in air with her hand*))

               -h. And the::n you wou::ld SPLIT them, let's say I wanted to split, spit them, split them by *twos.*

               One, two/

               OO|OO|OO|OO|OO

               And look. One, two, three, four, five.

               ((*taps chalk against each pair of circles*))

               THEN WHY DO YOU NOT CALL, -h, *ten* a, like-- a

               ((*Facing Sean, putting chalk back in tray*))

20 *Sean:*   ((*smiling*))

               //I disagree with myself. I call ten (an odd and even)]

21 *Mei:*   // an *odd* number] A::ND a even number? or why don't you call other, like numbers an odd number or-- and an even number?

22 *Sean:*   (3) I didn't think of that that way. Thank you for bringing it up, so:::

               I say it's:: (1).

               Five/ ten can be an odd and an even?

               //(???) can be an odd and a leven, an even]

In Episode 1, Sean merged the roles of dissenter and questioner to substantiate his disagreement with Sheena. Here, Mei has moved through the role of clarifier to dissenter, and finally, to dissenter-reasoner. She applies her established understanding that, in his statement about six, Sean is attending to *the number of groups of two* to generate a second example of an odd-and-even number, the instance of ten (Turn 19). Not only has she taken his perspective on the category of odd-and-even, but her example uses the same inscription that he used in his interaction with Tembe in Part 1, underscoring her strategy of appropriating his viewpoint in empathic dissent. From her central position at the board, she generates a second instance of an odd-and-even number using the same cookie representation. By reasoning from within his perspective, she sustains accountable argumentation in structuring this interaction. Unlike in the cross-representational discussion that took place in Episode 1, the students here are using the same inscription and thus able to have enough common ground in their argument to move it forward.

Figure 3.2. Sean and Mei disagreeing about six. Note Mei's drawing of ten, below Sean's drawing of six, on the chalkboard.
(Source: videotape of Deborah Ball's class, January 19, 1990, M.A.T.H. Project.)

The example of ten provides an even more significant move toward generalization: two examples of numbers with the same property (they can be partitioned into an odd number of pairs) open the possibility for a mathematically defined class of numbers. In her role as reasoner, Mei clearly generates the example of ten; however, because she is reasoning in service of her dissent, she does not interactionally take ownership of the example of ten. For her purposes, it is a strategic attempt to find a counterexample: a number that is known to be even but also has an odd number of groups of two. Because of her purposes in generating this example, she continues to position Sean as the originator of the example of ten in Turns 19 and 21 ("Why do you not call *ten* a, like-- a" "an *odd* number A::ND a *even* number?"). Although she has generated the second example of an odd-and-even, her statement continues in the second person (you) and thus maintains the category and the example as his. In doing so, she completes the revoicing move by then opening the slot for Sean to confirm or disconfirm her reformulation. After a three-second pause in Turn 22, Sean politely (if a little falteringly) revises his position to include both ten and six in the category of odd-and-even numbers.

Sean is active in the coconstruction of the odd-and-evens in this segment. From his position as principal, he asserts the boundaries of his category for six in Turn 18 by correcting Mei's restatement of his claim. By insisting that he is not claiming six is *odd,* but rather *odd-and-even,* he maintains the category of numbers under construction. Mei takes up the refined category in Turn 21 with an emphatic *"A::ND"* linking the emphasized words odd and even, underscoring that they are indeed talking about the same kind of number.

Sean thanks Mei for bringing up the example of ten (Turn 22), rejecting it as a counterexample and appropriating it as a second example using the resources of accountable argumentation. Accountable argumentation, with its expectation of modifying positions in light of convincing evidence, provides an interactional resource for Sean to "revise" his position without losing face (Renkema, 1993). He concedes to disagree with himself (Turn 20), revising his position, and allows ten as a second instance of an odd-and-even. Since he is still positioned to confirm or disconfirm this second reformulation, he maintains his role as principal of the controversy, as well as originator of the odd-and-even category.

Accountable argumentation sequences are primarily managed by the students engaged in them. Although these episodes were selected for close analysis in part because of Ball's peripherality, Ball nonetheless plays an important role in their execution. Although she is peripheral to the interaction, Ball does two kinds of work in this segment, one visible and one invisible. The visible work she does is to help Mei maintain the floor. By asking Riba to watch Mei, Ball transfers her teacher's authority to Mei, supporting her authority to speak. The invisible work that Ball does in this segment is one of non-intervention, related to the normative measured pace that supports accountable argumentation. Mei takes several pauses—one that lasts for 7 seconds (Turn 17)—while she is struggling to find her example of ten. Ball's silence and nonintervention allow Mei to do that challenging mathematical work.

Ball's choice of nonintervention during this 7-second pause is interesting to contrast with her decision to intervene after Sheena's 2-second pause in Episode 1. Recall that 2 seconds after Sheena starts to put the yardstick back in the tray, Ball steps in to close the argumentation sequence. During the pause in Episode 1, Sheena has signaled her readiness to relinquish her role as principal, whereas during Mei's 7-second pause, no such signal is provided. In fact, Ball has just signaled Riba to "watch what Mei is doing," recognizing that Mei is hard at work and should not be interrupted. This contrast highlights the subtle contextual cues that Ball relies on to support her students during accountable argumentation sequences.

Discussion of Episode 2: Through her roles as clarifier and then as dissenter-reasoner, Mei helps to move Sean's statement about six from a particular claim about six to a general class of numbers. Although further work by other class participants generates other examples of odd-and-evens, eventually mapping them onto a pattern on the number-line pattern, Mei's strategy of empathic dissent contributes to the mathematization of Sean's original statement. Empathic dissent, in contradistinction with the cross-representational talk of Episode 1, proves to be mathematically productive here and later in the class session.[6] Of course, Sean and Ball also contribute to the coconstruction of the concept during this episode, but Mei has done significant mathematical work.

The transformations of the statement about six can be summarized in Figure 3.3. In this figure, some important statements that highlight the construction of a generalization from Episode 2 are placed in chronological order. Mei first clarifies Sean's original statement, in the leftmost box. Her summary, in the second box, highlights his focus on the odd number of groups of two. The third box contains one of Mei's dissenting statements. In objecting, she names the source of the disagreement as "the number of groups," a more general description of a "type" of number. The fourth box contains the second instance of odd-and-even numbers, generated as a counterexample by Mei from her position as a dissenter. The last box includes a statement that alerts the class to the possibility that even more examples of such numbers may exist.

Sean:	S	**Mei:**	D	**Mei:**	D	**Mei:**	D	**Mei:**
Six can be an	U	I think what	I	It's not	I	Why do you	I	Why don't you
odd and an	M	he's saying is	S	according	S	not call 10 a,	S	call other like
even number.	M	that you have	S	to, like, how	S	like […] an odd	S	numbers an
	A	three groups of	E	many groups	E	number and a	E	odd […] and an
	R	2. And 3 is an	N	it is.	N	even number?	N	even number?
	Y	odd number.	T		T		T	
Particular								General

Figure 3.3. A map of the movement from the particular to the general through accountable argumentation in Episode 2.

Discussion

Revisiting the questions

Let us return to the questions that introduced this study: How can classroom discourse be organized to support mathematical disagreements that (a) are intellectually productive, and (b) minimize social discomfort? This analysis of the interactional organization of Deborah Ball's January 19, 1990, class provides one set of answers to those questions.

In this classroom, accountable argumentation brings the often hidden practices of mathematical reasoning into the visible world of classroom interactions; it allows for students to productively engage with the discipline (Engle & Conant, 2002). However, accountable argumentation entails something subtler than an

---

6 Whether Episode 1 could have become mathematically productive is an open question. My hunch is that more common ground would need to be established between Sean and Sheena, perhaps through the intervention of a clarifier who could help them link the different inscriptions they were using.

outright modeling of mathematical thinking. The social risks commonly associated with disagreement manage to be mitigated by other elements of its structure.

How does accountable argumentation make mathematical thinking a visible activity? First, the structure provided by accountable argumentation helps make disagreements intellectually productive in several ways: by providing support for thinking activities, by supporting deep engagement with specific ideas, and by supporting the learning and creating of new mathematics. Two norms of accountable argumentation, in particular, support thinking activities in the classroom: the use of terms from the mathematical and academic registers, such as *proving* and *conjecturing,* and the slow and measured pace of discussions that permit such activity to take place.

During the class session, these thinking activities are focused on a particular set of ideas. Accountable argumentation supports engagement with specific ideas, particularly through the expectations that *students attend to whole class discussions* and *students take a justified position in a discussion that they will act on or defend.* In addition, once they are engaged in a disagreement, the stakes for engagement increase. Dissenters are obliged to ask questions or otherwise substantiate their position to their peers. Principals, in contrast, must articulate their thinking to the whole class.

Effectively, these thinking activities and the engagement in particular ideas support both the learning and creation of mathematics. When Sheena has to articulate her position on zero, she transforms a vague statement to her teacher into a well-reasoned position to the whole class. This change can be viewed as increased competence and is indicative of her learning. Likewise, Sean "revises" his thinking about six as a unique "odd-and-even" number, recognizing that Mei's example of ten is also odd-and-even. He incorporates an example that had not been previously part of his thinking; he, too, has learned something new. Perhaps most striking, through the class' discussion, the particular observation about one number becomes the first of many examples of a general class of numbers—a counterexample is transformed into a second example through the process of argumentation, mirroring a way in which mathematics is created by mathematicians (Lakatos, 1976).

How do Ball's students tolerate the tension that is usually associated with disagreement in our culture? The discourse of accountable argumentation consistently distinguishes it from potentially threatening, personal disputes. By couching disagreements in academic terms, language provides not only a marker but a resource that allows students to challenge one another ("Prove it to us!"), make uncertain guesses (*conjecture*), and change their minds (*revise*) while avoiding the corresponding social costs of being deemed aggressive, foolish, or weak. Also, the deliberate civility of the exchanges (*please, thank you*) serves as a reminder that these arguments are not meant to be hostile. In addition, students sometimes deliberately disagree with positions ("what he said"), not *people,* again distancing the intellectual from the personal. Finally, disagreements are often initiated in the third person, changing to second person only after the principals have ratified their position.

Accountable argumentation provides subtle resources to mediate these tensions. The normalization of disagreements seems to play a significant role in alleviating potential risks. The students appear to know the norms, expectations, language, and roles through which to argue. Ball, through her talk, underscores disagreement as a collective, not personally threatening, activity through her use of the first person plural ("we're probably not going to settle this"). By helping students focus on the substance of one another's reasoning, accountable argumentation supports student learning through disagreement.

Limitations of this analysis: The importance of the teacher

In this study, I bring to the foreground the (mostly) student-mediated disagreements in this class session. I focus my analysis on two disagreement episodes in which the students themselves primarily manage the interactions. This strategy relegates the teacher to the background, underplaying her role in establishing and maintaining sustainable accountable argumentation. Yet even from this distance, the teacher's work emerges; an observer would be making a grave mistake to think that the students on their own can manage these disagreements because of the participation structure itself.

Within the disagreement episodes, the most obvious and ongoing pedagogical tension is whether the teacher should step in or stand back during students' discussions. Although an observer can sense Ball's grappling with this issue (as in Line 26 of the transcript), the data give us limited access to her judgments here. Other data—especially her journals and writings (e.g., Ball, 1998)—reveal the range of contextual knowledge employed in making these decisions throughout the class session. These resources include, but are not limited to, her assessments of the utility of students' exchanges for the whole class's learning, her plans for the day, and her sense of the "sturdiness" of individual students in the face of disagreements.

Furthermore, by focusing on one form of patterned interaction in the class's discourse, this analysis does not emphasize the many ways the teacher supports the viability of disagreements *throughout* the class session. First, in the strongly scaffolded disagreements that were not focused on here, Ball models the disagreement strategies explicated in this chapter by taking on roles, enacting norms, and using the academic registers and polite tones that she expects from her students. Even when she herself is not modeling argumentation practices, she exerts her authority from the sidelines of the student-student exchanges to assist them in sustaining their disagreements. She helps students maintain the conversational floor by focusing the audience's attention on their arguments. Even beyond all argumentation, she attends to the socio-emotional experience of disagreeing when, at the end of this particularly intense class session, she debriefs with her students about their experience of the controversies. For example, she asks the students (including Sean) how it feels to be the only person taking a particular position. This aspect of the work of teaching mathematics through disagreements is not highlighted in this analysis. Although the analysis of accountable argumentation might provide a starting point for teachers wanting to engage in complex teaching practices by providing images of, and language for, how students might learn through disagreements, it does not do justice to the nuanced and difficult judgments a teacher makes while orchestrating such discussions.

Acknowledgements

Support for this research was provided by UC—Berkeley's Spencer Center for Integrated Studies of Teaching and Learning. The thoughtful comments of Julia Aguirre, Rogers Hall, Cathy Kessel, Susan Magidson, Alan Schoenfeld, Natasha Speer, and Dan Zimmerlin and of the anonymous reviewers helped greatly in writing this chapter. Susan Jurow's feedback and conversation throughout the writing of this chapter were especially appreciated. Special thanks to Tamar Posner for her collaboration on the original analysis from which this chapter emerged. Portions of this analysis were presented at the 1999 annual meeting of the American Educational Research Association, Montreal, Quebec, Canada, and the 2000 Psychology in Mathematics Education conference, Cuernavaca, Morelos, Mexico. Please direct correspondence about this chapter to 115 Miller Hall, Box 353600, Seattle, WA 98195-3600, or send an e-mail message to lanihorn@u.washington.edu.

REFERENCES

Ball, D. L. (1998). *Crossing Boundaries: Probing the interplay of mathematics and pedagogy in elementary teaching.* Paper prepared for the Research Presession of the Annual Meeting of the National Council of Teachers of Mathematics. Washington, DC.

Ball, D. L., & Bass, H. (2000). Making believe: The collective construction of public mathematical knowledge in the elementary classroom. In D.C. Phillips (Ed.), *Constructivism in Education: Opinions and second opinions on controversial issues* (pp. 193–224). Chicago: University of Chicago Press.

Cazden, C. (2001). *Classroom discourse: The language of teaching and learning (2nd ed.).* Portsmouth, NH: Heinemann.

Chazan, D., & Ball, D. (1995). *Beyond exhortations not to tell: The teacher's role in discussion-intensive mathematics classes.* National Center for Research on Teacher Learning Craft Paper 95-2. East Lansing, MI.

Engle, R. A., & Conant, F. R. (2002). Guiding principles for fostering productive disciplinary engagement: Explaining an emergent argument in a community of learners classroom. *Cognition and Instruction, 20*(4), 399–483.

Glaser, B., & Strauss, A. (1967). *The discovery of grounded theory.* Chicago: Aldine.

Goffman, E. (1971). *Relations in public: Microstudies of the public order.* New York: Basic Books.

Goffman, E. (1974). *Frame analysis: An essay on the organization of experience.* New York: Harper & Row.

Goffman, E. (1981). *Forms of talk.* Philadelphia: University of Pennsylvania Press.

Goodwin, C., & Heritage, J. (1990). Conversation analysis. *Annual Review of Anthropology, 19,* 283–307.

Goodwin, M. H. (1991). *He-said-she-said: Talk as social organization among black children.* Bloomington, IN: Indiana University Press.

Greeno, J. G., Benke, G., Engle, R. A., Lachapelle, C., & Wiebe, M. (1998).Considering conceptual growth as change in discourse practices. In M. A. Gernsbacher & S. J. Derry (Eds.). *Proceedings* of the 20th annual conference of the Cognitive Science Society (pp. 442–447). Mahwah, NJ: Erlbaum.

Greeno, J., & Hall, R. P. (1997). Practicing representation: Learning with and about representational forms. *Phi Delta Kappan, 78* (5), 361–367.

Hanks, W. F. (1991). Foreword. In J. Lave & E. Wenger (Eds.), *Situated learning: Legitimate peripheral participation,* (pp. 13–24). New York: Cambridge University Press.

Hanks, W. F. (1996). *Language and communicative practices.* Boulder, CO: Westview Press.

Herrenkohl, L. R., & Guerra, M. R. (1998). Participant structures, scientific discourse, and student engagement in fourth grade. *Cognition and Instruction, 16,* 431–473.

Jurow, A. S. (2004). Generalizing in interaction: Middle school mathematics students making mathematical generalizations in a population-modeling project. *Mind, Culture, and Activity, 11*(4), 279–300.

Kazemi, E., & Stipek, D. (2001). Promoting conceptual thinking in four upper-elementary mathematics classrooms. *Elementary School Journal, 102*(1), 59–80.

Kieran, C., Forman, E. A., & Sfard, A. (2003). *Learning discourse: Discursive approaches to research in mathematics education.* New York: Springer.

Lakatos, I. (1976). *Proofs and refutations: The logic of mathematical discovery.* New York: Cambridge University Press.

Lampert, M. (1990, Spring). When the problem is not the question and the solution is not the answer: Mathematical knowing and teaching. *American Educational Research Journal, 27,* 19–63.

Lampert, M., & Blunk, M. (1998). *Talking mathematics in school: Studies of teaching and learning.* New York: Cambridge University Press.

Lampert, M., Rittenhouse, P., & Crumbaugh, C. (1996). Agreeing to disagree: Developing sociable mathematical discourse. In D. R. Olson & N. Torrance (Eds.), *The handbook of education and human development: New models of learning and teaching schooling* (pp. 731–764). London: Blackwell Publishers.

Lave, J., & Wenger, E. (1991). *Situated learning: Legitimate peripheral participation.* New York: Cambridge University Press.

Latour, B. (1987). *Science in action: How to follow scientists and engineers through society.* Cambridge, MA: Harvard University Press.

Lemke, J. L. (1990). *Talking science: Language, learning, and values.* Norwood, NJ: Ablex Publishing.

McClain, K., & Cobb, P. (2001). Supporting students' ability to reason about data. *Educational Studies in Mathematics, 45*(1–3), 103–129.

National Council of Teachers of Mathematics (NCTM). (2000). *Principles and standards for school mathematics.* Reston, VA: Author.

O'Connor, M. C. (1998). Language socialization in the mathematics classroom: Discourse practices and mathematical thinking. In M. Lampert & M. L. Blunk (Eds.), *Talking mathematics in school: Studies of teaching and learning* (pp. 17–55). New York: Cambridge University Press.

O'Connor, M. C., & Michaels, S. (1996). Shifting participant frameworks: Orchestrating thinking practices in group discussion. In D. Hicks (Ed.), *Discourse, learning, and schooling* (pp. 63–103). New York: Cambridge University Press.

Ochs, E. (1979) Transcription as theory. In E. Ochs & B. B. Schieffelin (Eds.), *Developmental pragmatics* (pp. 43–72). New York: Academic Press.

Philips, S. (1983/1993). *The invisible culture: Communication in the classroom and community on the Warm Springs Indian Reservation.* White Plains, NY: Longman.

Posner, T. (2007). From disagreement to understanding: Mathematical conversations in a classroom community. In *The study of teaching,* Monograph no. 14 of the *Journal for Research in Mathematics Education* (pp. 127–172). Reston, VA: National Council of Teachers of Mathematics.

Renkema, J. (1993). *Discourse studies.* Philadelphia: John Benjamins North America.

Resnick, L. B. (1999, June 16). Making American smarter. *Education Week, Century Series, 18*(40), 38–40.

Roth, W. M., & McGinn, M. K. (1998). Inscriptions: Toward a theory of representing as social practice. *Review of Educational Research, 68* (1), 35–59.

Sacks, H., Schegloff, E. A., & Jefferson, G. (1974). A simplest systematics for the organization of turn-taking for conversation. *Language, 50,* 696–735.

Schoenfeld, A. (1988, Spring). When good teaching leads to bad results: The disasters of "well-taught mathematics courses." *Educational Psychologist, 23,* 145–166.

Stigler, J.W., & Hiebert, J. (1999). *The teaching gap: Best ideas from the world's teachers for improving education in the classroom.* New York: Free Press.

Yackel, E., & Cobb, P. (1996). Sociomathematical norms, argumentation, and autonomy in mathematics. *Journal for Research in Mathematics Education, 27*(4), 458–477.

Chapter 4

Equity in a Mathematics Classroom: An Exploration

Tamar Posner
University of California at Berkeley

Over the past decade, reforms in mathematics education have called for "authentic" mathematical practices in the classroom. Recently, emphasis on equity in mathematics education has increased. These recommendations—authentic practices and equity—raise questions. How does one "see" equity in the classroom? How does one interpret authenticity? And are authentic mathematical practices and equity compatible?

Focus has shifted to the view that becoming a mathematician (or learning mathematics) does not simply mean acquiring a body of mathematical knowledge and skills, but rather acquiring "mathematical habits of mind" (Cuoco, 1998) or "mathematical dispositions" (Schoenfeld, 1992). Social practice theorists—and a growing number of educators—realize that mathematics is a social activity, a matter of beliefs, habits, and dispositions as well as skills (Cobb, 1996; Hall, 1996; Hall & Stevens, 1995; Lampert, 1997; Lave, 1988; Schoenfeld, 1992). If so, then—

> we may do well to conceive of mathematics education less as an instructional process (in the traditional sense of teaching specific, well-defined skills or items of knowledge), than as a socialization process. In this conception, people develop points of view and behavior patterns associated with gender roles, ethnic and familial cultures, and other socially defined traits. (Resnick, 1988, p. 58 as cited in Schoenfeld, 1992)

When mathematicians are engaged in actual processes of constructing mathematical knowledge, they make conjectures, test them against counterexamples or try to prove them, and revise their initial conjectures in an iterative process. Some mathematics educators have suggested that the goal of teaching mathematics should be "to bring the practice of knowing mathematics in school closer to what it means to know mathematics within the discipline" (Lampert, 1990; see also Ball, 1993, 1995; Cobb, Wood, & Yackel, 1991; Richards, 1991). One of the major ways to do so is to introduce students to the language and usage—the discourse—of the discipline. For example, the National Council of Teachers of Mathematics' *Curriculum and Evaluation Standards for School Mathematics* (1989) called for children to "talk mathematics" and for teachers to help them construct knowledge, learn to think in multiple ways about ideas, reflect on their own thinking, develop convincing arguments, and eventually extend the arguments to deductive proofs.

One result of this reform effort has been the suggestion that students construct and produce their own knowledge rather than receive "ready made" knowledge (NCTM, 2000).

Recently, a growing number of mathematics educators have recognized the need to move toward more equitable mathematics education (Ball, 2003; Linn & Kessel, 1996; Schoenfeld, 2002). Ball, for example, questions the relative silence of the mathematics education community. Echoing George Hein she asks, "How do concerns for equity play a role in the design of our efforts to improve mathematics and science instruction?" She suggests going beyond the rhetoric of "all students" or "high expectations" to address the "vast problems of educational inequalities that permeate U.S. schooling" (Ball, 2003).

Many reform initiatives reflect this trend. For example, *Principles and Standards for School Mathematics* (NCTM, 2000) includes a separate Equity Principle: "Excellence in mathematics education requires equity—high expectations and strong support for all students." Mathematicians and mathematics educators agree that "all students must have a solid grounding in mathematics to function effectively in today's world. The need to improve the learning of traditionally underserved groups of students is widely recognized; efforts to do so must continue" (Ball, Ferrini-Mundy, Kilpatrick, Milgram, Schmid, & Schaar, 2005, p. 2).

Although we hear a growing call for equity in mathematics education, we still need to explore what it looks like in practice. Schoenfeld notes that "like its antecedent [the NCTM 1989 *Standards* document], *Principles and Standards* can (despite its nearly 400 pages of densely packed text) be accused of being long on vision and somewhat short on detail. It identifies some essential goals, but does not provide a blueprint for achieving them" (2002, p. 15).

This chapter illustrates ways to look at how both equity and "authentic" mathematical practices play out in the daily interactions of a classroom. It examines some of the practices adapted from the mathematical community through minute-by-minute interactions in the classroom to better understand—and raise questions about—the social practice of mathematics and mathematics education. I share ways to examine and analyze interactions, providing a grounded framework for exploring classroom activity, to move us toward more equitable practices in the mathematics classroom.

THEORETICAL BACKGROUND

My examination of classroom interactions is guided by two theoretical perspectives: social practice theory and science studies. My analysis also draws on sociolinguistic tools aligned with those theoretical frameworks. Those frameworks are by no means independent of each other. Instead, they are "irreductionist" frameworks (Kaghan & Bowker, 2000) that account for individual agency as well as social structure and view a social order as continually produced and reproduced through ongoing practices and interactions.

Social Practice Theory

In this chapter, I use the concept of social practice to account for the complexity of human thought and actions as they take place in everyday life (Lave, 1988). The premise of social practice theory is that every activity—including the practice of school mathematics—is socially situated and occurs in a specific time and place, under particular historical and political conditions. (For example, the No Child Left Behind Act is a recent political condition that has had a direct influence on public school practices.) To understand social practice, one must look at what people do and say in their daily practices, and attend to politics and inequalities in resources and power.

Social practice theories highlight the interdependency of agency and structure, and seek to describe the dialectical relationship between them and understand the "interweaving of personal life and social structure" (Connell, 1987, p. 61). Those theories go beyond the dichotomy of structural determinism and psychological theories of agency. Understanding the relationship between social structure and individual actions is not a trivial matter. For example, to theorize this interdependency Giddens (1984) developed structuration theory, based on the idea that structure both influences and is influenced by human actions. In his view, actions have both intended and unintended consequences, and actors know much but not all about the structural ramifications of their actions.

In another attempt to theorize the relationship between agency and structure, Bourdieu (1977) introduced the concept of *habitus,* which is a system of dispositions, a set of acquired patterns of thought, behaviors, and tastes that constitute the link between social structures and human actions. For example, a child who grows up in a family and culture that strongly believes in creationism and goes to a school that emphasizes the theory of evolution might not easily understand the teacher's scientific discourse, might rely on different material to make meaning (Bible vs. scientific documents), and might feel alienated in terms of beliefs. The consequences of paying attention to those differences can be significant in terms of student participation in the classroom, and beyond.

Many studies indicate that children often experience difficulty in classrooms that are organized according to assumptions about the use of time, space, language, and instructional strategies that are different from those in their homes (Heath, 1983; Labov, 1970; Martin, 2000; McCollum, 1989; Philips, 1970; Willis, 1981). Although some studies attribute poor achievement to individual characteristics, social practice theorists closely examine the social and historical conditions under which students operate and have found other explanations. For example, Paul Willis studied working class boys in England. Through an examination of students' day-to-day interactions and the educational and social systems, he found that underachievement was related to their poor and working class background and was more often a result of rebellion against school authority than ability. Moreover, he argues that their rebellious behavior prepares them (sometimes consciously, sometime not) for working-class jobs. Willis found that working-class boys (the

"lads") tended to articulate a counter-school culture, which in its most basic dimension is "entrenched general and personalized opposition to 'authority'" (1981,
p. 11). The lads resisted with "a continuous scraping of chairs," "continuous fidg-
eting," "comic newspapers and nudes under half-lifted desks melt into elusive
textbooks," and more. In contrast, the conformists ("the ear'oles") "invest[ed] in
this formal structure, and in exchange for some loss of autonomy accept[ed] the
official guardians to keep the holy rules" (p. 22). They paid attention, their gaze
was focused on the teacher, and they did homework. Other studies have described
how African-American students are ridiculed by one another for "acting white,"
which is defined as behaviors characteristic of Caucasian students, for example,
getting good grades, liking classical music, raising their hands in class, and dress-
ing in a certain way. Social practice theorists argue that such day-to-day interactions contribute to the production and reproduction of social structures.

The foregoing two examples identify social trends and cannot be assumed true
for every member of a particular group. Although class, gender, or race may play
a role, individual agency cannot be neglected. For example, Martin (2000) studied
African American middle school students who succeeded in mathematics. Although
some of their peers viewed success in mathematics as "acting white," the students
studied had strategies that allowed them to simultaneously achieve academic success and social survival. Many maintained small friendship groups of "nerds" or
"good kids," and dismissed the ridicule of "bad kids." Grantham-Campbell (2005)
focused on successful Native Alaskan students and the cultural processes involved
in their success. She found that Native students who succeed in "doing school" are
able to suspend incidents of cultural conflict and maintain a calculated mistrust of
the schooling process. These same students are able to separate from the dominant
"image" of Natives and cultivate an "identity" drawn from positive experiences
and relationships. For example, when a student council calendar used a photograph
of Natives drinking, a Native student, in an act of defiance, dressed traditionally to
counter the negative representation of Native people.

Science Studies

To make sense of classroom interactions, I also draw on a body of work—science studies—that analyzes *science* as a social practice. This branch of science
studies focuses on examining the processes through which scientific knowledge is
produced. The approaches taken by science studies scholars vary, but to some degree, they all assume that scientific knowledge—as well as other knowledge—is
socially constructed—"made" collectively by many actors, both nonhuman and
human. Those researchers investigate how claims become "facts," and how and
why credit is allocated. I borrow these notions, and take the perspective that scientific "facts" are constructed by multiple actors.

Science studies researchers view scientific knowledge not as inevitable discoveries, but rather as productions of social, cultural, and material processes. Often
"scientific facts" are understood as claims about truth. Many science studies

scholars view such claims in a pragmatic way—meaning that a scientific claim is "true" if it "works" to the extent that relevant communities believe the claim. In that way "truth" is viewed as an historical construction, which can explain how knowledge is actually produced and claims become facts.

Early science studies focused on "science in the making." Through ethnographic-style fieldwork in laboratories, researchers made detailed accounts of the construction of scientific facts (Knorr-Cetina, 1981; Latour, 1979; Lynch, 1994) that described how scientific knowledge is accomplished through messy work. They found that successful scientific practice depends on multiple factors absent from positivist accounts of science. What emerged from these studies is that scientific facts are neither given nor discovered, but rather are outcomes of complex negotiation processes.

One of the first studies of this kind was *Laboratory Life* (1979). It describes the process by which scientific facts in a biology laboratory were gradually stripped of the conditions under which they were developed and became "black boxes"—claims whose validity and internal nature were not questioned. Latour argues (1986) that a claim alone is neither fact nor fiction; instead it is made so by other claims. Making a scientific fact includes gathering sufficient resources, enrolling or enlisting allies (human and nonhuman), and persuading others that the claim is a fact. Common ways to enroll allies (both human and nonhuman) are to cite other papers; use specific representations, such as graphs; drop names; and be associated with authoritative figures in the field. In short, the construction of a fact depends on interpretations and on who picks it up. Claims may involve dissent—a minority perspective that challenges a majority opinion. Dissent within science is expected. It plays dual roles: delegitimizing particular claims by exposing missing evidence or faulty thinking and legitimizing science as an institutionalized form of truth-seeking that evolves with new information.

Latour also describes the process through which a scientific statement goes through different modalities as it becomes a "hard fact"—from speculative hypothesis to proved statement to unspoken assumption. The modalities of a statement are modified through scrutiny in laboratories and conferences: how other scientists in the field cite it, certify it by assuming it is proved, and finally just assume that it is true. A scientific fact (black box) may spread to multiple communities yet have different meanings in each, becoming in Latour's words an "immutable mobile." Einstein's relativity theory as commonly understood is an example of an immutable mobile.

Assigning credit and invisible work

Steven Shapin (1989) documented instances of how 17th century chemist Francis Boyle assigned credit. Technicians were credited only when mistakes were made. When experiments went well, Boyle took full credit and the technicians were not mentioned. Other scholars have noted that an idea or a concept is rarely, if ever, developed solely by an individual and that making a fact includes erasure of elements that made it a "hard fact." Star (1991) calls these erased elements "invisible work." Once an idea is attributed to an individual, we often create a history in which other contributors become invisible and their contributions are erased (see, for example, Latour & Woolgar, 1979; Latour, 1987; Star, 1991a, 1991b; Strauss et al., 1985; Traweek, 1988).

Researchers have documented how work done by members of specific groups may become invisible: "rendering certain kinds of work invisible, reifying invisible things, and then secretly, privately, or duplicitously claiming the resources rightfully belonging to the work" and to the workers, who are often members of a marginalized group (Star, 1991, p. 279). This finding is consistent with findings of studies of women who enter such male-dominated professions as science and firefighting (Chetkovich, 1997; Eisenhart & Finkel, 1998; Ong, 2002; Traweek, 1988). For example, Chetkovich (1997) found that women firefighters were made invisible in different ways: excluded from conversations, spoken about as if they were not present, and denied credit for work they did in the firehouse. Classroom studies reveal that when females demonstrate competence in mathematics and science classrooms, teachers and male peers often refuse to acknowledge their achievements, and the girls themselves downplay their own ability. Moreover, they often deliberately position themselves to resolve such tension, usually by opting out of the conversation (Elkjaer, 1992; Ong, 2002, Tobias, 1990; Volman et al., 1999).

Sociolinguistic tools

Social practice theory—combined with science studies—offers productive theoretical orientations for exploring classroom discussion. Studies conducted by sociolinguists and educational ethnographers offer tools for analysis that are aligned with those frameworks. Sociolinguists consider power relationships to understand patterns in communication. For example, they look at utterances in conversations, patterns of utterances, and conditions under which utterances are produced, for example, getting the floor. I briefly outline some of their approaches, including the patterns of communication, discourse of disagreements, use of turns of talk, and register.

Patterns of communication: Everyday and classroom

Although I do not seek to perpetuate stereotypical behaviors, an important point to recognize is that communication in the reformed classroom may reproduce everyday patterns of communication. Lakoff (1975) argues that in general, females belittle and indicate doubt about the ideas they share more often than males do. Males are also seen as pursuing status by trying to win debates about ideas (Tannen, 1990; Tannen & Bly, 1993). In studies of classrooms, Noddings (1992) found that females' patterns of communication are less direct (and less aggressive) than males' patterns. Phelan (1993) talks about the ways females act incompetently and rely on the knowledge of boys who act competently in science classrooms.

These and other studies, for example, Cockburn (1985), Goffman (1977), Lie (1995), and Ong (2002), portray traditional gender roles in which females are depicted as generally supportive collaborators in classroom interactions. Conversely, males are generally portrayed as dominant individuals, obtaining, directing, and holding the conversational floor for extended periods.

A number of secondary school studies illustrate how gender influences the interactions in mathematics and science classrooms (Brophy, Guzzeti, & Williams, 1996; Kahle, 1990; Morse & Handley, 19985; Tobin & Garnett, 1987). Ong's study of undergraduate female minority students in physics explored the multiple strategies those students employed to "perform invisibility," including not taking responsibility for their own stances, bringing their own opinions in the names of others, and disengaging from discussions when they felt confronted in public. Lee (2004) also suggests that East Asian discourse is in general cooperative and harmonious, owing to its emphasis on "saving face."

Some of those findings have led to reforms intended to foster greater equity and close the achievement gap, such as sex-segregated mathematics and science programs in middle schools, and the creation of such programs as EQUALS. Those approaches are designed to restructure patterns of communication and increase the participation and engagement of typically nondominant participants.

Discourse of disagreements

Research on disagreements in everyday settings shows that often those who disagree try to mask their disagreement by restating, pausing, and self-repairing, for example, "um that, that one, [pause] um—zero has to be an odd, an even number." Moreover, the initial disagreeing party will often avoid using the term "disagree" because holding a disagreeing position can be uncomfortable (Davidson, 1984; Goodwin & Goodwin, 1987; Pomerantz, 1984).

However, public disagreements in a classroom setting during lessons on subject matter reveal a different pattern. Engle and Greeno (1994) differentiate *conceptual based* disagreements from *interpersonal* disagreements. In contrast with tentative and uncomfortable discourse displayed in interpersonal disagreements, participants in "conceptual based" disagreements quite eagerly engage in and announce their disagreements. Conceptual based disagreements follow the norm to "challenge ideas, not people," allowing safer participation—whereas interpersonal disagreements are mainly driven by personal, everyday reasons. Whereas conceptual based disagreements focus on explanations of concepts (in this chapter, on mathematical concepts), Engle and Greeno note that they also include social and interpersonal motivations that are managed in different ways than interpersonal "everyday" disagreements. They argue that participants in conceptually focused disagreements need to simultaneously satisfy multiple interpersonal and conceptual goals. For conceptual based conversation to be successful, participants must find ways to make their conceptual disagreements explicit while avoiding having it threaten the face-saving of the party they disagree with.

Lampert, Rittenhouse, and Crumbaugh (1996) have noted that the public airing, criticizing, and defending of ideas during public class discussions can potentially disrupt friendships and social relationships, especially if others lose face (Goffman, 1972). Furthermore, people may vary in their tolerance of, and comfort with, disagreement. Such variation may have its roots in *habitus*—individual dispositions or preferences of sociocultural groups. Important considerations in examining equity issues in the mathematics classroom are to discern the multiple reasons for engaging in public disagreements and to explore the habitus of the players.

Turns of talk

Everyday conversation is composed of speech between at least two people, organized by turns. The turn is the period of talk for each speaker. Ideally, one person speaks at a time; but this is not always so in a discussion. In formal situations, such as public lecture or rituals, turns of talk are often allocated by a facilitator or predetermined according to participants' roles. In an unstructured, spontaneous conversation, however, participants must determine in the moment when it is appropriate to take a turn. Sacks, Schegloff, and Jefferson (1974) suggest that certain rules govern the turn allocations in a conversation.

Classrooms may be unique because they are neither "formal" nor "informal." Sociolinguistic studies in education explore turns of talk and how they are allocated in the classroom. Verplae (2000) classifies the allocation of classroom talk into four distinct types: (1) the teacher's selection of a child who has not volunteered, (2) the student's response to the teacher's bid, (3) a student's request to speak while others are speaking (e. g., by raising his or her hand), and (4) a student's interruption of another speaker and the teacher's permission to that student to continue.

Philips describes different strategies used by students to take a turn of talk during classroom discussion. She also points out the unique situation in classroom discussions, in which the teacher plays an important role in allocating turns to talk. Schuman argues that a count of the number of turns taken by a participant is insufficient, rather one should ask how that student obtained the privilege to get the floor or, even more revealing, to take it.

A number of studies rely on turns of talk, length of turns, talk interruptions, and the teacher's nomination to the speaking floor to analyze how such issues as gender and race play into classroom interactions (Bousted, 1989; Kramarae & Treichler, 1990; Sadker & Sadker, 1990; Spender & Sarah, 1980). They found that students' cultural backgrounds, gender—in general what we might call their *habituses*—were correlated with the kinds of interactions they had in the classroom. These studies often adopt the perspective that the classroom is a microculture of the outside world, in which talk may form an important arena for the reproduction of gender, race, and class inequalities in human relations and social interaction (Baxter, 1999, p. 83).

A study by LaFrance showed that when compared with males in classroom discussion, females rarely are nominated to talk, talk less when they have the floor, and are interrupted more often (LaFrance, 1991). Similarly, Jones (1989) found that during classroom discussions, fewer females than males initiate questions or call out answers. Biggs and Edwards (1994) observed that teachers in multiethnic primary school classrooms interacted less frequently with children of color and that interactions with those students were less elaborate and shorter in duration.

Teachers are important figures in the classroom. Lampert (1990) notes, "The teacher has more power over how acts and utterances get interpreted, being in a position of social and intellectual authority, but these interpretations are finally the result of negotiation with students about how the activity is to be regarded" (pp. 34–35).

Register

The linguistic term *register* refers to the particular kind of language used in specific situational contexts. Particular kinds of activities require particular kinds of language. Often the nature of an activity can be determined by the style of language used during the activity. In this sense, language reflects the activity.

The mathematics register is made up of specific uses of language for mathematical purposes. It includes the words and syntax of spoken and written mathematics and their meanings. An implicit requirement to use language in certain ways exists in the mathematics classroom. Teachers introduce and model "mathematical" language. In part, learning school mathematics involves learning a specific register.

The school mathematics register has a specialized vocabulary and syntax. It contains both discipline-specific language (e.g., *isosceles, pi*) as well as words used in everyday language (e.g., *odd, even, complex*). For example, the word *show* in the context of a mathematics lesson often means to prove or justify an idea. However, in an everyday context, it can mean to "display" or "point out."

Although certain features of the mathematics register can be isolated and identified, the language used in mathematics classrooms cannot be regarded as a fixed or distinct set of words. Young students may sometimes conflate meanings of mathematical terms they learn in school with everyday meanings, or they may use two different meanings of a word in the same sentence. One example comes from a student in the lesson that is the subject of this chapter. Figure 4.1 shows two different meanings of the word *halves*, which she wrote in her journal.

In this description, "split in hafe" might have the everyday sense of two equal pieces of the whole (in this instance, the four circles), but the "hafes" refer to fractions—one of two equal pieces of a circle.

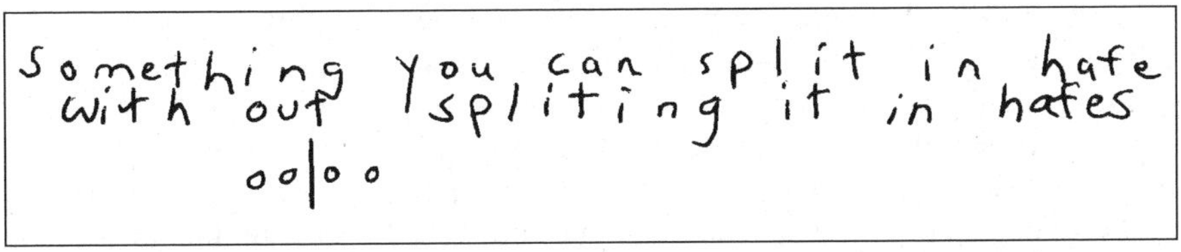

Something you can split in hafe
with out spliting it in hafes
o o | o o

Figure 4.1. A student's journal entry giving two different meanings of the word halves.

In this chapter, I consider school mathematics as a social practice in which teachers and learners use language to construct mathematical meaning. Mathematical meaning is constructed in part through specific linguistic practices associated with a mathematical register; moreover, learning mathematics is very much a matter of learning to speak using the appropriate register in the classroom.

METHODS

To understand the 6-minute segment in terms of classroom history, I grounded my findings with a larger-scale analysis of the whole lesson of January 19 and other supplemental data. By traversing past, present, and future accounts, I was able to view the 6-minute segment within a broader context.

In recent studies, educational researchers (e.g., Hall, 1996, 1999; Roth & McGinn, 1998) have focused attention on the material resources—such representations as the number line and classroom drawings, and such written materials as student journals—that mediate learning. I examined how classroom participants used talk and representations as they engaged in what seemed to be "mathematical" disagreements and tried to make sense of the classroom discussion.

To analyze the 6 minutes of video, I borrowed sociolinguistic methods. These include transcribing the episode with attention to details, such as how conversations were initiated, turns of talk, interruptions, speech repairs, and some attention to intonations, gestures, and duration of pauses (Goodwin & Heritage, 1990; Hall, 1996; McNeill, 1992; Sacks et al., 1974). Those methods aim to explicate how people produce interactions and what they accomplish in and through them, and can provide a lens to view the roles, social relationships, and power relationships among participants.

I employed a 6-step process. First, I used a detailed transcription developed by a colleague and myself (Posner & Horn, 1996) according to Ochs's (1979) conventions. In addition, I paid attention to the gestures, physical orientation, and material resources visible in the conversation, as well as to vocal elements. Although I use some different conventions in some transcript excerpts, for the convenience of the reader the line numbers of the transcripts correspond with those used in other chapters of this monograph. However, I have sometimes included additional details about classroom actions that help clarify the analysis. Second, to situate the given segment within a larger context, I examined participants' interactions within the 6 minutes in search of clues about classroom practices and history that contributed to the shape of the conversation. Third, I proceeded to find other examples of those practices in the rest of the available data. Fourth, I followed participants' utterances—as well as possible—into the recent past, because they were asked to reflect about the previous lesson. Examining transcripts from the available previous lessons provided some of the shared history in this classroom. Fifth, I focused on a detailed analysis of two sections of the 6-minute segment.

My sixth step involved an examination of previous lessons. In the midst of the analysis, I realized that to better understand the meaning participants made out of the conversation, I needed to explore their participation prior to this segment to sharpen my interpretation. The sixth step turned the analysis around because it allowed me to eliminate interpretations of events in the given segment that were erroneous owing to lack of context about the scope and sequence of mathematical conversation.

As much as possible, I tried to take into account the possible influences on classroom interactions. To do so, I relied on the research described previously on patterns of interactions in society, particularly in classrooms. One could argue that we do not have enough "evidence" to explain how such categories as race and gender play into the interactions discussed in this chapter. Nonetheless, failing to consider equity issues does not solve the problem. Research suggests that children are aware at all times of race and gender (Pollack, 2004). In particular, children from nondominant groups are aware if the person speaking is African American, Asian, or a girl. We have no reason to assume that this classroom is significantly different. Because race and gender are part of students' identities, part of their *habituses,* the power of those social dynamics is important to recognize.

However, I want to be clear that given the available data, my analysis is by no means an attempt to give an account of what "really" happened in this classroom. As the other chapters of this monograph indicate, the events can be interpreted in different ways.

Researchers have noted (Hall, 1999; Jordan & Henderson, 1995) that analyzing a segment of conversation without understanding its context can result in misinterpretation, even if we assume that participants in some instances are exhibiting the methods by which they carried out the activity. Moreover, by close analysis of talk in interactions out of context, we might end up with what Hall (1999) describes as "the crisis of interchangeability." That is, insufficient context may render the activities of different people indistinguishable.

The question of what constitutes enough context (Latour, 1987; Ortner, 1984) always lingers. The contextual boundaries I have drawn in this analysis are defined by the data available and the scope of the questions I chose to investigate. Although no definite answer can be given to what "actually" happened, my intention is to offer a possible, reasonable interpretation that takes into account participants' *habituses* (as afforded by the data available) and to lay the groundwork for a meaningful discussion of equity in mathematics classrooms. Rather than draw definitive conclusions from the data, I raise important questions to consider, particularly questions that have an impact on instructional practices. The intent of this analysis is to offer an interpretation of the 6-minute video segment that might be useful and relevant to other classroom situations, with the goal of moving us toward more equitable mathematics education.

DATA

The discussion in the 6-minute video segment analyzed is a continuation of an earlier discussion by a third-grade classroom on the nature of odd and even numbers on January 19, 1990. For a social practice theorist, the 6-minute segment lacks sufficient context to adequately analyze interactions with respect to equity. In addition to the 6-minute segment, however, I was able to view supplemental data sources, including three preceding mathematics lessons of this particular classroom (in which even and odd were discussed), the remainder of the lesson from which the 6-minute segment was extracted, journal entries of students and teacher (which include reflections on this particular lesson), and demographic information about the students.

Although the 6-minute segment and supplemental data sources provided a window into this classroom, I note that the video was recorded by an unknown person, under unknown conditions. Collecting data is always a subjective and selective activity (Goodwin, 1994; Hall, 2000). We highlight, include, and exclude according to our own perspectives and preferences. Those omissions and inclusions can lead to distortion and misinterpretation. This statement is true for any act of perceiving or recording data, including our own interpretive acts when we analyze those data.

An important consideration when addressing equity issues in the classroom is to explore how the demographics of the students and teacher might affect mathematical conversations (e.g., power dynamics, turns of talk, cultural preferences). Fortunately, those data were available. The demographics are shown in Appendix 2 to this monograph. As we can see in Appendix 2, this classroom contained 9 white and 10 black students, 9 male and 10 female students; 13 of the 19 students were proficient in the English language.

ANALYSIS

Classroom Culture

The mathematical pedagogy used in this classroom was intended to reflect practices of the mathematics community, as viewed by Lakatos (1976). The teacher describes this pedagogy as "based on a fallibilist epistemology" (Ball, 1988, p. 5). It adopts a "*particular perspective* on the nature of mathematical knowledge and activity" and "views the mathematical practice as a discourse community concerned with common questions and engaged collaboratively in pursuing and assessing mathematical ideas and in which more than traditional proof counts" (Ball, 1991, p. 46).

In this classroom, as in any classroom, were expectations that participants would behave and communicate in certain ways. As in the classroom depicted in *Proofs and Refutations* (Lakatos, 1976; see, e.g., pp. 76–77), claims, definitions, conjectures, and statements were attributed to individuals, and students and teacher referred to "Sheena's definition," "Mei's conjecture," and "Sean numbers." Students routinely made conjectures and claims, justified their positions, generated examples and counterexamples, transformed claims, and used the mathematical register. Teacher and students used a particular register, speaking of "conjectures," "proving," and "definitions."

Students and teacher used the words agree and disagree in reference to both people and ideas (see Table 4.1; see also Table 4.4 in this article's Appendix).

Table 4.1
Use of Agree, Agreement, Disagree, Disagreement in the January 19 Lesson

	With a person	With an idea	Unclear
Agree/agreement	9	11	5
Disagree/disagreement	8	9	6
Total	16	18	11

Previous three lessons

In the three previous lessons, the class discussed the nature of even and odd numbers. Sheena offered a definition, and Mei revised it. It became the "working definition" for even numbers (see Figure 4.2). Because the third graders could not come to a consensus about the evenness or oddness of zero, the teacher suggested that they organize a meeting on January 18 with the fourth graders to further discuss the matter. In that meeting, participants contemplated whether zero is even, odd, even and odd, or just a special number.

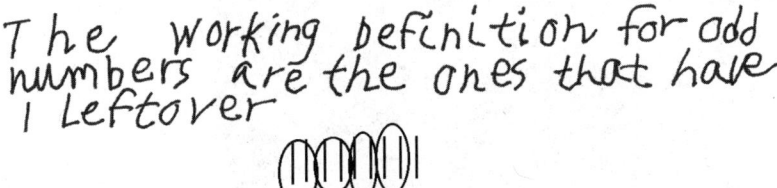

The working definition for even numbers is you can split the number and have the same amount of numbers on each side without using halves.

000/000

I would say the definition for an even number is, is two equal, two of the same numbers on each side without halfs.

The working definition for odd numbers are the ones that have 1 leftover

Well, an odd number is something that has one number left over ... After you circle the two's.

Figure 4.2. Definitions for even and odd in student journal entries and utterances.

During the January 19 lesson, the teacher repeatedly asked students to reflect on past lessons, particularly on the previous day's meeting with the fourth-grade class. During the January 19 class, "Ofala's definition" for odd number (see Figure 4.2) became another working definition.

The nature of even numbers was discussed on January 19. In particular the class discussed whether zero is even or odd. Several meanings of *even* were discussed, including definitions based on (a) the number line, (b) "two equal things make it," and (c) groupings of twos. Each of those meanings was associated with a particular kind of representation (see Figures 4.2 and 4.3). Distinctions related to meaning, definition, and property were not discussed.

A summary of the sequence of lessons, and of the class's working definitions, is shown in Table 4.2.

Analysis of the 6 Minutes on January 19

With this "history" as a backdrop, I next proceed with a detailed examination of two segments within the 6 minutes.

First segment: Is zero odd or even?

In the 6-minute segment, the teacher started the lesson by stating her agenda to solicit students' experiences from the previous day's discussion with the fourth graders. One of the students, Sheena, made the first attempt to answer the teacher's

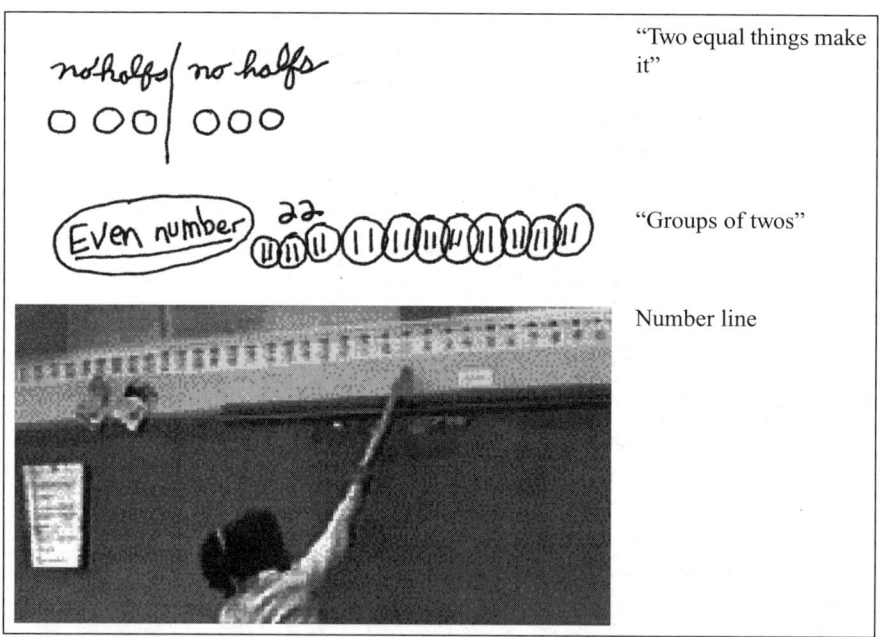

Figure 4.3. The representations associated with meanings for even.

Table 4.2
Occurrences in the Four Lessons

January 16	January 17	January 18	January 19
How to add and subtract numbers using the number line.	Discussion of the definitions of odd and even numbers (including Sheena's) posted on the chalkboard.	Meeting with the fourth graders.	Discussion of whether zero is even or odd.
Ofala offers a method to add numbers.	Betsy's method for comparing odd and even numbers, relying on the number line.	Three different ideas about zero are discussed: Zero is a special number.	Sheena says that zero has to be even, based on the number-line pattern.
Sheena's definition of even number, revised by Mei.	Disagreements between Sheena and Sean.	Zero is even and odd: "Well, I think zero is even and odd because nobody can really prove that it's even or odd."	Sean disagrees.
	Discussion of whether zero is even or odd.	Zero is even, argument based on the number line.	Nathan thinks that zero is special.
	Sean argues that one is even throughout the lesson. Other students disagree.	The nature of even and odd numbers is discussed at length.	Sean says that six could be even and odd.
	Lucy offers a definition of *even* based on the number line.	Discussion of whether zero is nothing or a number.	Discussion of what Sean means, contributions from several students.
	Lucy offers a conjecture: if a number ends with an odd, it is odd; with an even, it is even.	Discussion of the "use" of odd and even numbers. Argument that if all numbers would be even, we would not need a distinction between even and odds.	After the lesson, the teacher writes in her log about "Sean's discovery."
	Betsy says that on the basis of the number line, zero has to be odd.		
	Sean: zero is not a number, extended argument with him. Teacher asks students to stop arguing with Sean.		

Note. Events of the 6 minutes on January 19 are indicated in the box in the rightmost column.

request. Sheena described how the previous day's discussion with the fourth graders had helped her better understand the nature of zero. The teacher then asked her to give a particular example of how she revised her understanding, and the following exchange occurred:

			Transcript commentary
6[1]	Sheena:	Well, I didn't think that zero was—zero, um—even or odd until yesterday. They said that *it could be even* because of the ones on each side is odd, so that *couldn't be odd.* So that helped me understand it.	*Evidence that Sheena appears to under-stand odd and even as mutually exclusive. She refers the alternation of even and odd on the number line.*
7	Teacher:	Hmm. So y— So you thought about something that came up in the meeting that you hadn't thought about before? Okay.	
8	Sheena:	(*nods*)	
9	Teacher:	Other people's comments? Sean?	

Sheena tells the teacher how what other people ("they") said in the previous day's meeting helped her understand how zero "could be even" and "couldn't be odd" on the number line (6)[1]. Then the teacher reiterated what Sheena said about how the previous day's discussion influenced her thinking. They both appear satisfied with this exchange (7, 8). Then the teacher solicited additional comments from other students (9).

Instead of providing the teacher with comments regarding his own experience, Sean continued to discuss Sheena's ideas, and engaged her in the following disagreement:

			Transcript commentary
10	Sean:	Um, I—I—I just want to say something to Sheena, when sh— what she said about um that, that one, um—zero has to be an odd, an even number bec— I disagree because, um, because what what two things can you put together to make it?	*Sean changes the teacher's agenda. He refers to the "two things make it" definition.*
11	Sheena:	Could you repeat what you said, please?	
13	Sean:	Okay, um, I disagree with you because, um, if it was an even number, how— what two things could make it?	*Note that Sean dis-agrees with "you" rather than "your idea."*
14	Sheena:	Well, I could show you it. [Moves toward the chalkboard and points to the number line above the chalkboard.] Um, I forgot what his name was—but yesterday he said that this one [points to the 1 on the number line] and each—this one is odd and this one [points to the –1 on the num-ber line] is odd, so this one has to be even.	*Again, Sheena appears to understand odd and even as mutually exclusive.*

[1] The number indicates line number of the transcript.

15 Sean:	But, that doesn't mean it always is even.	*Does Sean think that even is a changeable property?*
16 Sheena:	It could be even.	*Sheena softens her claim from "has to" to "could."*
17 Sean:	It could be, but . . .	
18 Sheena:	I'm not saying that is has to be even. I meant that it could be.	*Sheena maintains the change from "has to" to "could."*
19 Sean:	You said it was.	

Through this exchange, Sheena seemed to change her position a number of times. In the beginning, she stated that zero could be even and couldn't be odd (6). After a short exchange with Sean, she concluded that "zero has to be even" (14). Yet few turns later, she stated that what she meant was that "it [zero] could be [even]."

Using the multiple lenses of social practice theory, science studies, and sociolinguistics, we ask, Did Sheena change her understanding about the nature of zero through the work of the disagreement? Was she convinced by Sean's argument? Did she understand the difference between claims for necessity and claims for possibility? Was she confused? Or simply holding a space for what others had previously said?

Both Sheena and Sean provided justifications for their positions. Although Sheena's position seemed to have changed, the justifications she gave for both positions were similar. Both heavily depended on alternating patterns on the number line. In contrast, Sean's justification, which was given in the form of a question (10, 13), suggests that he used a different definition of even numbers, in which even numbers are made out of two things. (We have yet to determine what the "things" are.) Did Sean and Sheena use different representations differently to think about even numbers and, as a result, misunderstand each other? Was representation the core of their disagreement? Was their disagreement conceptually based or interpersonal or both?

Sheena justified her position(s) on the basis of what a third party had argued the previous day. This third party was not present in the classroom during this discussion. That party at first was referred to by Sheena as a general "they" and at a later utterance became more specifically "Um, I forgot what his name was—but yesterday he said" (14). Bringing an explanation in the name of somebody else can be interpreted in multiple ways, from using the other one as an ally to strengthen a claim (Latour, 1987) to a strategy for avoiding personal responsibility in the event that the claim turns out to be "wrong" (Ong, 2002), to the documentation of a historical event. At this point, the reason Sheena brought in the fourth grader as the principal originator of her claim is unclear.

After performing two self-repairs, Sean reiterated Sheena's position inaccurately, stating that she had said zero has to be even whereas she said zero could be even (6). What was Sean disagreeing with? Was he disagreeing with what Sheena actually said (e.g., zero could be even and cannot be odd)? Or was he disagreeing with what he said she said (e.g., zero has to be even)? Sheena may have been wondering about those same questions as she asked Sean to repeat himself (11). A clear mark of disagreement was drawn by Sean's opening statement (10), which were repeated in his subsequent turn (13).

Sean was clear and up-front about his disagreeing status, and repeatedly marked himself as the disagreeing party (10, 13). As Engle and Greeno's research suggests, perhaps for Sean the disagreement was primarily a "conceptual based" one about the nature of zero rather than an "interpersonal" one. However, as the conversation continued, his motivations for disagreeing with Sheena were unclear and might have changed. After the short exchange between Sean and Sheena, Sean challenged Sheena and argued that her explanation did not "mean it [zero] always is [even]" (15). Sheena did not object and replied that "it could be," to which Sean seemed to agree that "it could be but ..." (17). At this point, one could expect the disagreement to dissolve, because the engaging parties had reached an agreement that zero could be even. Yet Sean continued, "but ..." (17), to which Sheena quickly replied, "I'm not saying that is has to be even. I meant that it could" (18). Her intonation suggested that she expected to end this exchange having stated her agreement with Sean (e.g., zero could be even). Moreover, in addition to clarifying her current position, "I'm not saying that it has to be even," in accordance with the classroom practices, she emphasized her meaning, insinuating that even in the past (if she had said differently) she "meant it [zero] could be [even]" (17). He replied quickly, "You said it was" (18). At this point one wonders if the issue at stake for Sean was not whether zero had to be or could be even (which they seemed to agree about); rather it was whether Sheena was right or wrong.

To better understand the meanings Sheena and Sean made out of this exchange, I returned to the previous lessons in which I searched for visible patterns of Sheena's and Sean's behavior as well as their contributions leading to the current discussion. With these brief "histories" at hand, we are better situated to consider the previous questions:

Did Sheena change her understanding about the nature of zero? Was she convinced by Sean's argument? Did she understand the difference between claims for necessity and claims for possibility? Was she confused?

Lessons of January 16, 17, 18, and 19: Sheena

Sheena is an African American girl. She is a native English speaker and is a relatively new member in this classroom (three months, see Table 4.1). On several occasions in the previous lessons, she attributed the justification of her positions to third parties (mainly boys) who were not present during her talk.

Sheena seemed to have a clear grasp of the nature of even and odd numbers. In a previous lesson she provided a verbal definition of an even number: "I'd say that the definition for an even number is um, a number that you can split." She gave six as an example, "Say you have six, so I'll make this ... and then you want to have it so you can split it in half and so you split what you have, the same amount of numbers on each side." She made a drawing on the chalkboard (see Figure 4.4).

Figure 4.4. Sheena's representation for six.

She defined even numbers as numbers that are "made out of two things" (other numbers). Later, her classmate, Mei, revised Sheena's definition.

> But, I, I don't think they ... yeah, but I still don't think that you could, that you could do it. Well I think what Sheena, I think Sheena should revise to that, even numbers that have the, numbers that you can split that have the same amount on each side without having to have halves.

This definition was discussed at length and referred to as "Sheena's definition" nine times in the three previous lessons. Sheena often used tentative, uncommitted language (e.g., could, might). She did not occupy the public floor often. During the January 17 and 19 lesson, she had 13 and 14 turns (see Table 4.3), and when she did, her turns were mainly allocated to her by the teacher.

Lessons of January 16, 17, 18, and 19: Sean

Sean is a Caucasian boy. He is a native English speaker and was one of the "old timers" in the classroom (two years). Sean was an active participant in classroom discussions and frequently occupied the classroom floor (see Table 4.4). He seemed to be a dissenter; when asked by the teacher later in the January 19 lesson if he was comfortable in that position, he replied that he was.

Teacher: What about you, Sean? A couple of times this week, you've had, you've taken a position that nobody else in the class agreed with. What does that make, does that make you change your mind, how does that make you feel?

Sean: It makes me feel fine.

He was the initiator of many disagreements and frequently used the word *disagree* to mark his position (see Table 4.4). At one point in the January 17 lesson, the teacher explicitly asked students not to further argue with Sean (see Table 4.3 for further details). On other occasions, some students explicitly refrained from arguing with him. Sean repeatedly used self-repairs, pauses, and unfinished sentences in his talk. As a consequence, other classroom participants (mostly the teacher) often reiterated his utterances in an attempt to clarify his intent. These reiterations usually included sensible interpretations, which seemed distinctly different from his original sentences. They were often posed in the form of questions that were longer and more detailed than Sean's original utterances. Sean seldom credited definitions, claims, or ideas to other participants. To some extent Sean appeared to grasp the classroom's definition of even numbers (as one can see from the way he couched his initial question to Sheena). However, his understanding of this concept seemed unstable, in that he frequently made errors in his attempt to classify numbers as even or odd. For example, he argued that 30 is odd because

Table 4.3.

Number of Turns Each Participant Had During the January 17 and 19 Lessons

Name	Number of turns	
	January 17	January 19
Teacher	184	208
Sean	50	114
Ofala	7	28
Mei	12	27
Tembe	3	21
Riba	8	20
Cassandra	5	20
Betsy	12	16
Sheena	13	14
Nathan	4	10
Jeannie	10	8
Keith	5	7
Lucy	15	1
Maria	—	1
Lindiwe	13	0
Daniel	2	0
Mark	7	0
Devin	—	0
Harooun	2	0
Tory	7	0
Unknown	104	0

Note. The January 17 statistics are compiled from a transcript provided by Ball. In that transcript 104 turns were attributed to "student" or had no attribution. The content of many of the turns attributed to "student" suggests that they may have been Sean's. Statistics from the January 16 and 18 transcripts are not included in this table because many turns of talk are not attributed to particular students.

3 is odd (Teacher's log, 1990). Later in the January 19 class, he said that an odd number was one that could be split into an odd number of groups (415), probably meaning an odd number of groups of two. On a few occasions, to justify his position, he turned to "personal" justifications, such as, "One is even because my mom said so... ." He believed her "because I trust her."

Sean often had a disproportionately large number of turns to talk compared with the other students in the classroom (see Table 4.3). For example, the lesson from which the analyzed segment is taken comprised a total of 521 turns of talk. The teacher had 208 turns, and Sean had 114 turns. The next most frequent student speaker was Ofala, with 28 turns of talk. The extensive floor time Sean had in the classroom is also reflected in this chapter (i.e., I was able to provide more history for Sean than Sheena because of more available data). Turns were both given to him by the teacher and initiated by him. When competing for floor time with other students, he was often the winner. When he got the floor, on several occasions he did not answer the question the teacher asked the class. Rather he had "something else" to say and changed the teacher's agenda. (For example, see lines 9–10 and lines 59–60.)

59 *Teacher:* More comments about the meeting?

(*Mei whispers:* Oh I have. *Her voice rises at the end. She raises her hand.*)

I'd really like to hear from as many people as possible what comments you had or reactions you had to being in that meeting yesterday.

Sean?

60 *Sean:* (*Sean is off camera, Mei's hand is still up.*) Um, I don't have anything about the meeting yesterday, but I was just thinking about six, that it's a . . . I'm just thinking. I'm just thinking it can be an odd number, too, 'cause there could be two, four, six, and two, three twos, that'd make six ...

Returning to the analysis of January 19

Sheena's utterances within the context of the previous lessons suggest that her understanding of even numbers was stable. Moreover, she was one of the "originators" of what was called in class the working definition of even numbers (see Table 4.3). In addition, she had offered a visual representation of this definition on the classroom chalkboard. Thus one might reasonably assume that Sheena was well versed in the "two things make it" definition of even numbers and was comfortable and familiar with its visual representation. Yet rather than join Sean in the representation he used—a representation that she was familiar with—as he challenged her position, she chose the number line to illustrate the explanation for her position that zero is an even number. Even though this choice warrants an explanation, the available data do not allow us to resolve this issue.

Not all representations are created equal. Sheena chose for her justification to show the pattern on the number line. The number line was a permanent participant in the classroom (Latour, 1987). In fact, it had quite an honorable and visible position in front of the classroom above the chalkboard; this position suggested it was endorsed by classroom authority (e.g., the teacher). In comparison, "two things make it" was of a different status. Its visual representation was a temporary participant in the classroom, and it often disappeared as the chalkboard was erased. In addition, it was introduced by a female minority student, Sheena, to support a definition of even number she proposed. Thus it may not have had the authoritative power of the durable number line. Not surprisingly, Sheena chose the number line (a durable representation) over the flimsiness of the "two things make it" representation as an ally in her argument with Sean. The number line's authoritative stance might explain her persistence in using it, and the fact that she refused to join Sean in his use of an alternative definition even though she appeared to understand that definition well. Sean, in contrast, kept on using the "two things make it" definition from the previous classroom discussions. Ironically, Sean used Sheena's definition to make his point, whereas she relied on the authoritative number line.

As Hall (1999) reminds us, the challenge of analyzing a conversation is that to some degree we are entering in the middle. That challenge was certainly confronted in this analysis, because the classroom had a history that includes discussion of the same topics as those discussed in the analyzed segment. On the one hand, Sean may have been referring to the class's working definition of even numbers, and perhaps was holding Sheena accountable to this definition, which she had introduced and which had been named after her. On the other hand, given his "history," we might reasonably suspect that for Sean, alternating between definitions and representations was difficult; he appeared confused about the meaning of *even*. Confusion may have been the reason he kept on referring to the "two things make it" definition and refused to accept Sheena's explanation based on the number line. Only toward the end of the exchange did he shift from challenging her position that zero could be even to challenging what she actually said.

Still other matters remain to be explained. If Sheena was solid in her understanding of the nature of even numbers, why did she change her mind through the short duration of this exchange, from saying that "zero could be even" to saying that "zero has to be even" and back to saying she meant to say that "zero could be even"?

In the analyzed segment, Sheena twice voiced her position in the name of a third party that was not present at the time of the discussion (e.g., "they said," "he said"). This tactic, combined with her use of tentative language (a common style preference among females, as discussed in the "Theoretical Framework" section) might indicate that Sheena, to some degree, did not want to be held responsible for her position.

Sheena appeared reluctant to take verbal authorship of her position. Yet when Sean repeated his challenge to her, she used more forceful language, which was compatible with her explanation. However, as the disagreement continued, Sheena backed off. In an attempt to dissolve the disagreement, she returned to the use of noncommittal language, saying, "I meant it could be even," to which Sean quickly replied, "but you said it was," ending the disagreement with a possible confrontation—eventually showing she was "wrong." At the point in which Sheena backed off, she chose to actively disengage from this disagreement, and in some ways she contributed to her own invisibility.

Second Segment

Whereas the first segment showed the interwoven nature of conceptually based and interpersonal disagreements, the second segment illustrates possible uses of everyday and mathematical meanings for *even* in the same utterances. In this analysis, I use even with its common mathematical meaning, that is, an integer is even if it can be written as $2n$ for some integer n.

After the first segment, the teacher revisited her agenda and asked the students, once again, for their comments about the previous day's meeting with the fourth graders. Nathan, in response to her request, shared his experience. Nathan (in accordance with classroom practice) stated that as a result of the meeting, he revised his idea from thinking that zero was even to thinking that zero was special. To justify his position that zero is a special number and not an even number, Nathan strove to "make up" the set of even numbers in such a way that zero was not, and could not be, part of the made up set, thereby to prove his point that zero was a special number. In the following transcript, I indicate two constructions that Nathan appeared to make to construct the even numbers. Nathan responded to the teacher's request with the following statement:

Transcript commentary

25 Um:::, first I said that um (.) zero wa::s e::ven but then
I guess I revised so that zero, I think, is special because
um, I, um,
::: marks indicates lengthened syllable; each : equals
one beat.
even numbers? Like they they make even numbers; like
two, um, two makes four, and four is an even number
and four makes eight eight is an even number and u::m,
like that.

Intonation drops.   CONSTRUCTION ATTEMPT 1
                                        Drop in intonation suggests
                                        that Nathan has completed
                                        his first attempt to con-
                                        struct the even numbers.

And, and go on like that. and like one plus one and go   CONSTRUCTION ATTEMPT 2
on adding the same numbers with the same numbers.
And so I, I think zero's special.

I started with the assumption that Nathan's argument made sense (McDermott, 1993). Nathan repeatedly used the terms *make, made,* and *go on,* which are terms that often indicate a mode of construction (Walkerdine, 1988). Attempt 1 has two possible interpretations. One is that Nathan was trying to construct the evenly evens, the even numbers greater than 2, that is, 4, 8, 12, 16, … , using 2 + 2, 4 + 4, 6 + 6, and so on. A second interpretation is that Nathan was trying to construct numbers by doubling, beginning with 2, and resulting in 4, 8, 16, 32, … , using 2 + 2, 4 + 4, 8 + 8, and so on. Whichever of these constructions was intended, he still had a problem, assuming that he was trying to construct the set of all even numbers. Neither of his constructed sets included all the even numbers he needed to justify his position that zero was special. His constructed sets still had many "missing evens." After Nathan's first attempt to construct the even numbers, his intonation drops as he utters, "like that," indicating an ending of a train of thought or a conversation. Given that Attempt 1 did not result in the desired set, his making Attempt 2 is not surprising. This time he took the same numbers and added them to make "another" even number. He says, "and like one plus one and go on adding the same numbers with the same numbers."

This method of constructing the set of even numbers is consistent with the classroom's working definition of even numbers as numbers made by two equal things. This method expanded Nathan's initial set so that all positive even numbers (excluding zero) were accounted for. Hence, as far as he was concerned, he had offered a reasonable justification of his position that zero was special, and his turn was over. However, for the teacher it was not the end. She needed some clarification. After promising the class to return to her agenda (comments on yesterday's meeting), she asked Nathan, "Were you saying that when you put even numbers together, you get another even number?" (26). She may be asking whether Nathan means that the sum of two even numbers is even. He gives a short reply, "Yeah" (27).

The teacher continued to question Nathan about his intentions, asking, "Or were you saying that *all* even numbers are made up of even numbers?" (28) After a short hesitation, he replies once again, "Yeah, they are"[2] (29).

The mathematical meaning of the word *or* is inclusive, but its everyday meaning is exclusive. The teacher's use of *or* at the beginning of her second question to Nathan might have indicated (to Nathan and the rest of the class) some problem in compatibility between the two choices. However, Nathan still answered with a positive gesture, implying that for him both options were viable—meaning both statements were correct:

> "When you put even numbers together you get an even number," *and* "All even numbers are made up of other even numbers."

To help make sense of this dialogue, what follows is some of Nathan's "history."

Lessons of January 16, 17, 18, and 19: Nathan

Nathan is an Ethiopian boy. He is a fluent English speaker and was one of the "old timers" in the school—three years (see Table 4.1). As seen in Table 4.2,

---

[2] This transcription differs from the other used in this monograph. Readers may wish to watch the accompanying video and decide for themselves.

Nathan said little during the lesson. His lack of participation had also been evident in the previous three lessons, in which he had a total of five turns. The teacher commented in her journal that he did homework incorrectly owing to what she thought might be a misinterpretation of the task. For Nathan the word *even* had multiple significations. He often used *even* to mean "equal" and "same." At other times he used *even* with its usual mathematical meaning.

Returning to analysis of January 19

In this dialogue Nathan and the teacher may have been using the term *even* differently. *Even* has multiple meanings, depending on the context in which it is used (Walkerdine, 1988). It could be used in multiple ways across participants, across utterances of the same participant, or even within an utterance (Goodwin, 1994; Walkerdine, 1988). The word *even* has been used in this particular classroom in different contexts. Often *even* meant a number that is in the set of even numbers. On other occasions, *even* was used as part of "everyday language" and meant "equal" or "the same." Thus in the context of this particular classroom, the utterance "made out of two even numbers" could have different meanings: (1) numbers that belong to the set of even numbers, such as four and six; or (2) two equal numbers, such as three and three. It is possible that for Nathan, the two alternatives presented by the teacher posed no problem, since "when you put two even numbers together, you get an even number" *and* "all even numbers are made out of even [equal, same] numbers." For Nathan, perhaps both statements made sense with the classroom's working definition.

The teacher continued to question Nathan's meaning. She revoiced his utterances, this time including both statements, combining them with *but,* once again indicating the existence of some contradictions. In the transcript that follows, I annotate instances in which different meanings of *even* allow different interpretations of utterances.

			Transcript commentary
34	Teacher	Nathan said a minute ago that when you put even numbers together you get an even number,	*Teacher revoices what Nathan said. Possible ambiguity: "Sums of equal numbers are even" or "sums of even numbers are even."*
35	Betsy	Mm-hm.	
36	Teacher	*But he also* said, I think? (.) that *all* even numbers are made up of other even numbers	*Possible ambiguity: "All even numbers are made out of even numbers" or*
		(.) indicates very slight pause.	*"All even numbers are made out of equal numbers"*

As they continued (in lines 37–43), both the teacher and Nathan used the same words, *even, same,* and *equal.* However, those words may have had different meanings for both participants at different parts of the conversation. On the one hand, the teacher used *even* to refer only to numbers belonging to the set of even numbers. For Nathan, on the other hand, the word *even* had multiple significations. He often used even to mean "equal" and "same," and at other times used even to include numbers belonging to the set of even numbers.

The teacher, in her attempt to clarify Nathan's utterances (34), made the distinction of the evenly even: "when you put even numbers together you get an even number." Betsy (35) seems to agree with that statement.

In the next part of the discussion, the teacher says that she thinks Nathan also said that all even numbers are made out of even numbers.

			Transcript commentary
37	Mei:	I disagree. (*raises her hand*)	*Mei may be disagreeing with "all even numbers are made out of even numbers."*
38	Sheena:	(*says something to Mei*)	
39	Teacher:	Two even numbers just the same.	*Ambiguous: "two equal numbers that are just the same" or "two even numbers that are just the same."*
40	Nathan:	Unh-uh. (*Nathan presses pen in palm of left hand*)	*It is not clear what Nathan is agreeing with.*
41	Teacher:	The same even number?	*The teacher is not sure what Nathan is agreeing with and attempts to clarify. Possible ambiguity (as lines 34a, 34c): The same equal numbers (e.g., three and three) or the same even numbers (e.g., four and four).*
42	Nathan:	Yeah, like four.	*Nathan agrees.*
43	Teacher:	Like eight is four plus four? Are all the even numbers— can you do that with all the even numbers? That they'd be made up of two identical even numbers?	*Possible ambiguity: Are they made up of two identical equal numbers (creating the full set of even numbers) or two identical even numbers (creating a subset of the even numbers, the evenly even numbers[3])*

In lines 45–58, Nathan and Betsy talk further with each other; Betsy tries to make sense of Nathan's statements, and seems to realize that Nathan is talking about a subset of the even numbers (the evenly even numbers). Mei and Sean interject comments (indicated in gray in the transcript that follows). Betsy and Nathan face each other, appearing not to attend to Mei's and Sean's comments, but rather to be engaged in their own conversation.

---

[3] The set of evenly even numbers is obtained by successive doubling: 1, 2, 4, 8, 16, 32, 64, and so on. This set is discussed in more detail at the end of this section.

			Transcript commentary
45	Betsy:	Not—not—Like six. Six is //two, two, ...] (*Betsy turns back around in her seat facing Nathan and shaking her head.*) // *indicates beginning of overlapping utterances.] indicates end of overlapping utterances*	*Betsy understands the statement as "all even numbers are made out of even numbers" and provides Nathan with a counterexample: six is made out of two odd numbers (part of the oddly evens).*
46	Sean:	//Six is two odd] (*off-camera*) numbers to //make an even, to make an even number.] (*His utterance overlaps Betsy's next utterance, line 48.*)	*Sean's utterance begins toward the end of Betsy's utterance.*
46a	Betsy:	//Six you can't get two] (*facing Nathan*)	*Betsy continues to talk with Nathan (indicated by the fact she is facing him).*
47	Mei:	Three and three// (*Mei indicates three and three with her fingers.*) *Mei's utterance overlaps Betsy's, line 50.*	*Mei echoes Betsy's counterexample.*
48	Betsy:	//THREE TWOS make six. You can't put a four and a four (.) or a....] *Capital letters indicate increased volume.*	*Betsy continues to address Nathan, emphasizing that three twos make six—meaning that not all even numbers are made out of two equal even numbers.*
48a	Nathan:	Yeah but... (waving pen)	*To which Nathan agrees but has something more to say.*
49	Sean:	Three twos??? *??? indicates rise in intonation — a questioning tone.*	*Sean does not appear to be part of the conversation between Betsy and Nathan.*
50	Betsy:	THREE'S—Three is odd. (*Betsy continues to face Nathan.*)	
51	Sean:	Or, um—	
52	Nathan:	I know that, but um, um I'm talking about like, two plus two is four, and four plus four is eight and I just skipped the six so, I just added the ones that, that add. Like the two plus two is four, and four is an even number and I'm just talking about (.) the things that um, like—	*Nathan replies to Betsy (about the fact that six is made out of two odds) and emphasizes that he is only talking about a subset of the even numbers—the evenly evens—stressing that he skipped the six.*

53	Sean:	…. Six can be an odd number.	*Possibly picking up on Betsy's comment.*
54	Nathan:	what I just said—the um:: (3),um like two is plus two is four and four plus four is eight //and— *3 indicates a 3-second (relatively long) pause*	
55	Betsy:	//So what] you're doing is you're going by twos //and] *In the next few turns Betsy's and Nathan's utterances overlap.*	*In an attempt to understand what Nathan is saying, Betsy reconstructs the set (the evenly evens) that Nathan refers to.*
56	Nathan:	//Yeah]	*He agrees.*
56a	Betsy:	then what two equals from then you go from—all the way //up]	*Betsy continues to discuss "Nathan's set"—the evenly evens—and indicates the evenly evens are constructed according to a general pattern.*
56b	Nathan:	//Yeah], I'm not going by EVERY SINGLE NUMBER. Like, *// indicates his talk is overlapping Betsy's. Capital letters indicate increased volume.*	*Nathan agrees with Betsy and emphasizes that he is talking only about a subset of the even numbers.*
57	Betsy:		*Okay.*
58	Nathan:	two=four=six=eight. *= indicates there are no gaps between utterances.*	*Finishing his statement in line 61 and indicating that he is not including all the even numbers: like 2, 4, 6, 8. Instead he is talking about a subset, even numbers that are sums of the same.*

After this exchange, the teacher returned to her agenda, asking for more comments about the meeting.

			Transcript commentary
59	Teacher:	(4) More comments about the meeting? I'd really like to hear from as many people as possible (.) (off-camera)	*Returning to her previous agenda.*
59a	Sean:	Oh! (off-camera)	
59b	Teacher:	what comments you had or reactions you had to being in that meeting yesterday.	*Teacher finishes statement in line 64.*
59c	Mei:	Oh I have (??) (raises her hand as she whispers)	*Mei seems eager to get the floor— she moves in her seat as she raises her hand to get the teacher's attention.*
59d	Teacher:	Sean?	*Sean gets the floor.*

The camera shows Mei with her hand raised, answering the teacher's request for comments about the meeting. Sean also responded and got the floor. In line 60 he says that he does not have a comment about the meeting but has something else to say, indicating that he was aware that he was not answering the teacher's request, but changing her agenda.

			Transcript commentary
60	Sean:	Um, I don't, I don't have anything about the meeting (.) yesterday, but I was just thinking about six, it was a . . . I'm thinking that it's a, it's an o:dd, it can be an odd number, too, 'cause there could be (.) two, two, two=four=six. (.) Two, three twos, that'd make six ... (*Sean is off-camera. Mei's hand still raised.*)	*Sean changes the teacher's agenda from discussion of yesterday's meeting to discussion of six being an odd number.* *Sean states that he is thinking of the number six (not a set of numbers in contrast with the set discussed by Nathan and Betsy) and that six is made of two threes and three twos. His "two, four, six" is similar to Nathan's "two, four, six" in line 61.*
61	Teacher:	Uh-huh . . .	
62	Sean:	And, and two threes,	
62a	Teacher:	Mm//hm]	
62b	Sean:	// it] could be an odd and an even number. (2) Be both. (1) Three things to make it //and]	*As Sean and other students (e.g., Betsy) noticed before, six can be made of two odd numbers. However, unlike the other students he insists that six is both odd and even.*
62c	Teacher:	// Uh] huh	
62d	Sean:	there could be two things to make it.	*Sean is still talking about the number six (not a set of numbers).*
63	Teacher	And the TWO things that you put together to make it were ODD, //right?] *Capital letters indicate increased volume.*	*The teacher asks Sean if the "two things" are odd.*
63a	Sean	// Uh]-huh	
63b	Teacher	Three and three are each odd?	*The teacher asks Sean again if the "two things" are odd.*
64	Sean	Uh huh, and the thing, the OTHER, those two/ the TWOS (.) we::re (.) even.	

Sean argued that six could be odd and even. His justification echoed Betsy's observation about six (48, 50): Six is made of three twos and three is odd. However, Betsy did not conclude from her observation that six is odd.

In lines 65–67, the teacher connects Sean's and Nathan's comments.

			Transcript commentary
65	Teacher	So you're kind of/ I think Nathan said then that he wasn't talking about every even number, right, Nathan?	*The teacher connects Sean's argument that the number six is odd because it's made of three groups of two with Nathan's comment about a set (the evenly evens). Her comment suggests that she was aware of Nathan's idea about a subset of even numbers that doesn't include all even numbers.*
66	Nathan	Yeah, //I'm not] (*Pen in mouth.*)	*Nathan confirms that he was not talking about all even numbers, but rather a proper (i.e. doesn't contain all even numbers) subset of the even numbers.*
67	Teacher	//Were you] saying that? Some of the even numbers (.) like six are made up of two odds, like you just suggested.	*The teacher appears to be talking to Sean. She revoices his statement about six being odd because it is made up of two threes as a more general statement that "some of the even numbers like six are made up of two odds."*

The teacher started by talking to Sean. In line 65, she connected Sean's comment about a single number with Nathan's construction of a set of numbers by referring to what Nathan said. Then, in line 67, she implied that Sean's suggestion that "six is odd because it's made of two threes" is an example of a more general phenomenon, the oddly evens.

Later in the January 19 lesson

Later in the lesson, the idea of six being odd and even was discussed further. Several times during the lesson students responded to Sean's talk about six by saying that if a number could be decomposed into an odd number of groups of two, that it was not necessarily odd and even itself. For example, Keith said, "That doesn't necessarily mean that six is odd" (75), and explained further, "Because, just because two odd numbers add up to an even number doesn't mean it has to be odd" (78). Later, Riba also disagreed with Sean (315) and said, "I'm going to show . . . see, it doesn't . . . um, it doesn't matter how much circles there are, it matters when you like put one, two, three, four, five, six . . . it doesn't matter about, it doesn't mean . . . how much times you circle two, it doesn't prove that six is an odd number" (317).

Mei tried to express what she thought Sean was saying, "I think what he is saying is that it's almost, see, I think what he's saying is that you have three groups of two. And three is an odd number so six can be an odd number *and* a even number" (145). After the teacher asked her if she agreed with that (six being odd and even), Mei said that she disagreed (149) and gave an example (153),

Let's say that I have [pauses]. Let's see. If you call six an odd number, why don't [pause] let's see [pause] let's say ten. One, two . . . [draws circles on board] and here are ten circles. And then you would split them, let's say I wanted to split, spit them, split them by twos. . . One, two, . . . , one, two, three, four, five . . . [she draws]

$$\text{o o|o o|o o|o o|o o}$$

Sean thanked Mei, then said, "so—I say it's—ten can be an odd and an even" (158). Later, Ofala gave another example, "Well, I just think, um, that just because you put twenty-two there and that makes eleven, that that doesn't mean that it's an odd number. My conjecture, I think it's always true, because if all twos are circled in a number that's even, then it's an even number" (431).

SUMMARY OF IDEAS, WORDS, AND REPRESENTATIONS USED IN THE DISCUSSION

Figure 4.5 provides a summary of pivotal elements in the conversation. The first two rows show Sheena's definition of January 17 as revised by Mei, and the remaining rows refer to the January 19 lesson. The first column tracks the use of the number line and such related ideas as skipping and patterns of numbers. At the beginning of the lesson Sheena used the number line to illustrate her idea about zero being even. Much later in the lesson Riba used the number line to describe the pattern that she saw. The second column of Figure 4.5 tracks the use of the "two things make it" definition and its associated representation. The third column tracks the use of grouping by twos and its associated representation. The fourth column tracks participants' word use. At the beginning of the lesson, Sheena used the phrases *has to be,* which she subsequently softened to *could be.* Later, students talked about "can be odd" and "can be odd and even." Throughout the lesson, *even, same,* and *equal* were used interchangeably. Note that the representation is used to discuss sets of numbers and patterns of numbers, but the representations associated with "two things make it" and "groups of two" are not.

Discussion of the characteristics of odd and even numbers can be traced to at least 500 AD. In book I of his *De arithmetica,* Boethius identified three subsets of the even numbers, the "evenly-odd," "evenly-even," and "oddly-even" (Swetz, 1995). Two of these subsets (evenly odd and evenly even) emerged in this classroom discussion.

When an evenly even number is divided into two equal parts, the parts are always evenly even and can again be separated into two evenly even parts, until the process ends with one. Such numbers are obtained by successive doubling: 1, 2, 4, 8, 16, 32, 64, and so on. Nathan's remarks appear to be an attempt toward constructing this set.

Evenly odd numbers are those that can be divided into equal parts, but the halves cannot again be separated into equal parts. Examples are 2, 6, 10, 18, 22.

	Skipping, patterns 5 4 3 2 1 0 1 2 3 4	"Two things make it" OOO/OOO	"Groups of two"	"Could be," "can be," "has to be" "Same," "equal," "even"
Sheena Jan. 17		Two things make it . . .		
Mei Jan. 17		without using halves		
Sheena 14	Pattern of odd and even on the number line			Zero has to be even.
Sheena 16, 18				Zero could be even.
Nathan 25				"[E]ven numbers make even numbers" . . . "adding the same numbers with the same numbers."
Teacher 28				"[E]ven numbers are made out of even numbers?"
Teacher 39 to 43				"Two even numbers just the same." "The same even numbers?" "Identical even numbers?"
Betsy 48, 50			Three twos make six. Three is odd.	
Nathan 52, 56b, 58	"and I just skipped the six" "[N]ot going by every single number like two, four, six, eight"	"[T]wo plus two is four, and four plus plus is eight."		
Sean 60, 140	[T]wo, four, six	o|o|oo|oo	Three twos make 6.	[S]ix can be odd.

Figure 4.5. Ideas, words, and representations in the air.

Speaker	Skipping, patterns (← 5 4 3 2 1 0 1 2 3 4)	"Two things make it" (OOO/OOO)	"Groups of two"	"Could be," "can be," "has to be" "Same," "equal," "even"
Mei 145				"I think what [Sean]'s saying is that you have three groups of two. And three is an odd number so six can be an odd number and a even number."
Mei 153–155, 157		o|o|o o|o o|oo		If you call six odd . . . then why don't you call ten like . . . an odd and an even?
Sean 162	"I'm just skipping over to the even."			"It wouldn't be all odd and even. . . . I'm just skipping over to the even."
Riba 390			Twenty doesn't work.	
Sean 398				Twenty-two can be an odd number.
Riba 401				How come 24 can't be one?
Teacher 407	Is there a pattern?			
Sean 409	It's every four.			
Riba 417 to 428	Shows "every four" on the number line		[T]wo works, one is odd.	"I think Sean's saying that some even numbers in a pattern can be even and odd and some can't."

Figure 4.5 (continued). Ideas, words, and representations in the air.

Evenly odd numbers are generated by multiplying odd numbers by two, and the successive numbers in the evenly odd sequence differ by four. Sean's remarks were revoiced by the teacher (e.g., lines 67, 341), Mei, and later Riba (see Figure 4.5) in ways that moved the class closer to constructing this set.

DISCUSSION

In the segments analyzed, we see two discussions that were cultivated by the teacher to publicly explore a mathematical concept in the classroom. Although both were about the understanding of odd and even numbers, they had some essential differences. This section is organized around three main issues illuminated through my analysis: (1) the shift from conceptually based to interpersonal disagreements, (2) the use of everyday and mathematics registers in the classroom, and (3) the practice of assigning competence and attributing "discoveries" to individuals. I remind the reader that the data available permit me to illustrate how equity considerations can be seen through an analysis of daily interactions in a classroom, but that this analysis can only raise questions about what happened in this particular classroom.

Shift from Conceptual to Personal Disagreements

The first segment showed how content-based talk can morph into interpersonal talk. This shift can have an important effect on the goals of students and teacher. When such a shift occurs, we see that participants either retreat from the conversation or preserve some posture that does not necessarily have much to do with the content at hand.

How might such shifts be affected by *habitus?* Previous studies indicate that gender and race tend to affect student participation (e.g., girls retreat from confrontation, nonnative speakers are more reluctant to speak out, etc.). When Sheena and Sean's conversation shifted from conceptual to personal (e.g., you said this, I said that), Sheena withdrew from the conversation and barely participated thereafter. Although we do not have enough information to understand why she withdrew, Sheena's withdrawal raises an issue that teachers deal with daily in classrooms. Without teacher intervention, classroom participants might also implicitly or explicitly use those backgrounds as tools to position themselves in the classroom. This dynamic might contribute to problems with equity within the mathematics classroom.

Such shifts could become fertile moments for teacher intervention, were a teacher to redirect the conversation back to the content. Looking at such moments can be a way for teachers to see how close examination of talk—and attending to such shifts—can inform them in refining their classroom practice. In the given segment (and the overall lesson), on a number of occasions students and teacher expressed their disagreements by naming people rather than the idea itself (e.g., "I disagree with Sheena" rather than "I disagree with Sheena's definition"; see Table 4.5 in this chapter's Appendix). This kind of naming might also contribute to some students' shying away from expressing their ideas (Ong, 2002) and to some students' challenging the ideas of others. In such instances, teachers have an opportunity to redirect the conversation by framing a student's idea as a concept (e.g., "So you disagree that zero is odd?") rather than attribute it solely to an individual.

If we can find ways to locate those moments of shift in talk as they happen, then intervention can be fruitful. Through our intervening and keeping the conversation focused on concepts, classrooms might become safer spaces for students who traditionally do not speak out in discussions. Thus, identifying and intervening in such moments can become a practical way to address equity. Further investigation is warranted in understanding and sharing best intervention practices.

Use of Everyday and Mathematical Registers

In this classroom, *even* had multiple meanings, including "equal," "made of the same two things," and "member of the set of even numbers." My analysis suggests that the different meanings for *even* in this classroom contributed to difficulties in communication among the participants and possibly led to confusion about the nature of "even" numbers. Whereas mathematicians are more fluent and consistent in the use of mathematical language, young students typically use everyday and mathematical registers interchangeably, a tendency that can lead to misunderstandings based on their comprehension of terms used in everyday life. In fact, they sometimes use a mathematical term with different meanings in the same sentence (Schegloff, 1992).

We need to be sensitive while engaging in mathematical conversations with students, especially those who are not fluent English speakers. Possibly, for example, Nathan was misunderstood because of the way he used the word *even*. Teachers need to know that "everyday" and "mathematical" registers can be the basis of miscommunication; probing further to clarify terms can be helpful in fostering understanding.

Attribution

One of the practices used by this classroom teacher has been called (by the teacher herself) "assigning competence" (Ball, Goffney, & Bass, 2005, p. 4; Boaler, 2003). This practice involves teachers' raising the status of students that may or may not be of a lower status in a group. To do so, for example, the teacher might praise something a student has said or done that has intellectual value, and bring it to the group's attention; ask a student to present an idea; or publicly praise a student's work in a whole-class setting. One way in which teachers do so is by revoicing students' ideas in their own words.

O'Connor and Michael (1996) note that "revoicing" is a powerful strategy teachers use to facilitate discussions and scaffold student understandings. It is also a means by which teachers give validation and credit for ideas to students who may not have been recognized by their classmates. By aligning students with particular ideas, revoicing can also position students as thinkers and hypothesizers. I would argue that the subtle difference between revoicing and inventing a student idea can also be unclear and open to interpretation. In this example, from the comment "six is even and odd," the class moved to the classification of numbers as "oddly even" through a series of revoicings and translations (see e.g., transcript lines 67, 145, 334, 341, 417).

After the lesson, the teacher wrote in her journal (Ball, 1993, p. 387),

> I'm wondering if I should introduce to the class the idea that Sean has identified (discovered) a new category of numbers—those that have the property he has noted. We could name them after him. Or maybe this is silly—will just confuse them since its nonstandard knowledge—i.e., not part of the wider mathematical community's shared knowledge. I have to think about this. It has the potential to enhance what kids are thinking about "definition" and its role, nature, and purpose in mathematical activity and discourse, which, after all, has been a major point this week. What should a definition do? Why is it needed?

The following day, the teacher returned to the classroom and named the oddly even numbers the "Sean numbers." "He seemed pretty pleased," she wrote in her journal "and the others surprised." There might be multiple reasons for her decision, from simply stating what she thought had happened to a social consideration of Sean's position in the class. Regardless of her intentions, on which I do not wish to speculate at length, I cannot help but wonder, how did Sheena feel? How did Mei feel?

Sean had by far the most turns of any student to talk during the lesson (see Table 4.4) and many turns during the previous three lessons, including turns given by the teacher or other participants or taken by himself. The amount of time his voice occupied the public floor might have contributed to the impression that he was the initiator of, or the main contributor to, the emerging concept. I argue that naming the concept "Sean number" makes invisible the contributions of the other students who coconstructed it. It could have just as easily been attributed to Mei or Riba (see Figure 4.5).

My analysis suggests that several students, including Sheena, Nathan, Betsy, Mei, and Riba, contributed work that became invisible with the making of the "fact" of Sean numbers. Were they pleased with the idea of Sean numbers too? Did they also assume that Sean "discovered" the idea? Did they recognize the teacher's action as an attempt to raise Sean's status by assigning competence? After all, his utterances in discussions may often have seemed unclear and were revoiced by other students. Or might they have attributed her naming the concept after Sean to their own invisibility, thus to some degree erasing their contributions to its development?

"Sean numbers" had consequences in the classroom. "[O]ver the course of the next few days, some children explored patterns with Sean numbers, just as others were investigating patterns with even and odd numbers" (Ball, 1993, p. 187).

CONCLUDING REMARKS

Analysis of classroom situations like this gives us the opportunity to acknowledge that many (if not all) ideas in the classroom—and in mathematical communities—are coconstructed. Indeed, in Ball's classroom, coconstruction was valued and encouraged. Reflections, justifications, and inquiry played major roles in learning and knowledge construction, and were held as core values in the classroom community.

Moreover, this teacher practiced these values, not just among the students in her classroom but in the larger research and teaching community. In a reflexive act, aligned with her classroom practices, she opened her classroom to the eyes of many different communities, allowing, and even encouraging, various interpretations of her own teaching. This courageous act provides researchers with a valuable opportunity to analyze a classroom episode from different perspectives and with different tools. In addition, it provides teachers with an opportunity to think about how different pedagogical strategies can ensure equitable participation. In particular, the segments analyzed in this chapter suggest questions about the affordances of different classroom participation structures and different mathematical representations.

From the perspective of social practice, "Sean numbers" cannot be seen only in terms of the teacher's intentions, because its unintended consequences (Giddens, 1979) went beyond the walls of this specific classroom. From the perspective of science studies, what matters is who, and under what conditions, took the attribution of the concept to Sean to be true (Latour, 1986). This "lesson" to some extent became an "immutable mobile," and traveled far and across communities for over a decade, from the mathematics education community to the teacher education community and even to the mathematics community. It is now known as "Sean numbers," which was to some degree "black boxed" (Latour, 1987) as the "outcome" of the lesson. This "black boxing" erases the history of how the comment "six can be even and odd" became the hard fact of the "Sean numbers." Because this lesson is often used as an example of a mathematics lesson in teacher education programs across the country, one wonders about the unintended messages sent to those, primarily female, teachers. Is this example perpetuating the image of a sole male discoverer of mathematical knowledge? Was Sean's visibility achieved at the expense of other students' invisibility? Does it reinforce the invisibility of female contributions in mathematics? Did other students, such as Mei or Riba, get too little credit by not being named?

Lampert and others argue that the goal of teaching mathematics should be "to bring the practice of knowing mathematics in school closer to what it means to know mathematics within the discipline" (Lampert, 1990). In this respect, participants in this classroom engaged in practices resembling those of the mathematical community. At this point a word of caution seems necessary. Although I agree with this goal, I also agree with Ball (1993, p. 377) that certain aspects of the discipline, such as "competitiveness among research mathematicians—

competitiveness for individual recognition, for resources, and for prestige—is hardly a desirable model for an elementary classroom." We need to be selective about the mathematical practices we introduce into the classroom, because some practices may perpetuate inequity.

Attributing a "discovery" to an individual and naming it after that individual is not necessarily a desirable practice for an elementary classroom, even though it is done among mathematicians. In the scientific and mathematical communities, the processes of making "facts" are often erased, removed from "mathematics in the making," leaving a fact stripped of its history. Those processes might contribute to the impression that few women are mathematicians, and that even fewer have contributed in a significant way. Creating and naming a "fact" are political processes, and their examination can help us better understand how mathematics has been documented (Bowker & Star, 1999).

Not all mathematical practices are desirable, and not all are equitable. More equitable practices in elementary mathematics classrooms have two possible benefits: an immediate one for students in that classroom, and the possibility of those students' growing up to become members that help the mathematical community become more equitable. We recognize the importance of equity in society as a whole, and in the classroom in particular. However, detecting equity is difficult. This chapter presents examples of ways to analyze equity in day-to-day classroom interactions. Once inequity is detectable, our work to address equity issues can begin.

ACKNOWLEDGMENTS

As I emphasize in my chapter, no work is produced solely by an individual. This statement is definitely true in the development of this chapter. Cathy Kessel was instrumental in shaping and editing my drafts, and her skill and spirit are greatly appreciated. Thoughtful comments of Maria Ong, Natasha Schull, Tony Torralba, and the anonymous reviewers also contributed to this chapter. I especially thank Alan Schoenfeld for his insight and support, and Deborah Ball, who graciously shared her classroom and logs with a wide audience. I began working on this analysis several years ago in collaboration with Ilana Horn. We had many fruitful discussions and debates, which influenced my analysis; thank you, Lani.

This chapter is dedicated to my dear friend and colleague, Mary Grantham-Campbell, who provided her own perspective on equity and relentlessly advocated for all children. She was the first person to review this chapter and, sadly, passed away before its publication.

REFERENCES

Ball, D. (1988). *Knowledge and reasoning in mathematical pedagogy: Examining what prospective teachers bring to teacher education.* Unpublished doctoral dissertation, Michigan State University, East Lansing.

Ball, D. (1991). Research on teaching mathematics: Making subject-matter knowledge part of the equation. In J. Brophy (Ed.), *Advances in research on teaching* (Vol. 2, pp. 1–48). Greenwich, CT: JAI Press.

Ball, D. L. (1993). With an eye on the mathematical horizon: Dilemmas of teaching elementary school mathematics. *Elementary School Journal, 93*(4), 373–397.

Ball, D. L. (1997). January 16, 1990). Teacher's journal.

Ball, D. L. (2003) Reflections by Deborah Loewenberg Ball, TERC, 2003.

Ball, D. L., Ferrini-Mundy, J., Kilpatrick, J., Milgram, J., Schmid, W., & Schaar, R. (2005). Reaching for common ground in K–12 mathematics education. *Notices of the American Mathematical Society, 52*(9), 1055–1058.

Ball, D. L., Goffney, I., & Bass, H. (2005). The role of mathematics instruction in building a socially just and diverse democracy. *Mathematics Educator, 15*(1), 2–6.

Baxter, J. A. (1999). Teaching girls to speak out: The female voice in public contexts. *Language and Education, 13*(2), 81–98.

Biggs, A., & Edwards, V. (1994). "I treat them all the same": teacher-pupil talk in multi-ethnic classrooms. In D. Graddol, J. Maybin, & B. Steirer (Eds.), *Language and literacy in social context* (pp. 82–99). Clevedon, England: Multilingual Matters.

Boaler, J. (2003). When learning no longer matters: Standardized testing and the creation of inequality. *Phi Delta Kappan, 84*(7), 502–506.

Bourdieu, P. (1977). *Outline of a theory of practice.* Cambridge: Cambridge University Press.

Bousted, M. (1989). Who talks? *English in Education, 23*(3), 41–51.

Bowker, G., & Star, S. L. (1999). *Sorting things out: Classification and its consequences.* Cambridge, MA: MIT Press.

Chetkovich, C. (1997). *Real heat: Gender and race in the urban fire service.* New Brunswick, NJ: Rutger University Press.

Cobb, P. (1996). Accounting for mathematical learning in the social context of the classroom. In C. Alsina, J. M. Alvarez, B. Hodgson, C. Laborde, & A. Perez (Eds.), *8th International Congress of Mathematics Education selected lectures* (pp. 85–99). Seville: S.A.E.M.

Cobb, P., Wood, T., & Yackel, E. T. (1991). A constructivist approach to second grade mathematics. In E. von Glasersfeld (Ed.), *Radical constructivism in mathematics education* (pp. 157–176). Dordrecht, Netherlands: Kluwer.

Cockburn, C. (1985). *Machinery of dominance: Men, women and technical know-how.* London: Pluto.

Connell, R W. (1987). *Gender and power: Society, the person and sexual politics.* Cambridge: Polity Press.

Cuoco, A. (1998). Mathematics as a way of thinking about things. In *High school mathematics at work* (pp. 102–106). Washington, DC: National Academy Press.

Davidson, J. (1984). Subsequent versions of invitations, offers, requests, and proposals dealing with potential or actual rejection. In J. Heritage & J.M. Atkinson (Eds.), *Structures of social action: Studies in conversation analysis* (pp. 102–128). Cambridge, England: Cambridge University Press.

Eisenhart, M., & Finkel, E. (1998). *Women's science learning and succeeding from the margins.* Chicago: University of Chicago Press.

Elkjaer, B. (1992). Girls and information technology in Denmark: An account of socially constructed problem. *Gender and Education 1*(2), 25–40.

Engle, R. A., & Greeno, J. G. (1994). Managing disagreement in intellectual conversations: Coordinating interpersonal and conceptual concerns in the collaborative construction of mathematical explanations. In *Proceedings of the Sixteenth Annual Conference of the Cognitive Science Society.* Hillsdale, NJ: Erlbaum.

Giddens, A. (1979). *Central problems in social theory: Action, structure, and contradiction in social analysis.* Berkeley: University of California Press.

Giddens, A. (1984). *The constitution of society.* Cambridge, England: Polity Press.

Goffman, E. (1977). The arrangement between sexes. *Theory and Society, 4,* 301–331.

Goodwin, C. (1994). Professional vision. *American Anthropologist, 96,* 606–633.

Goodwin, C., & Heritage, J. (1990). Conversation analysis. *Annual Review of Anthropology, 19,* 283–307.

Goodwin, M. H., & Goodwin, C. (1987). Children arguing. In S. Philips, S. Steele, & C. Tanz (Eds.), *Language, gender and sex in comparative perspective* (pp. 200–248). Cambridge: Cambridge University Press.

Grantham-Campbell, M. D. (2005) *Recasting Alaska Native students: Success, failure and identity*. Unpublished doctoral dissertation. Stanford University.

Hall, R. (1996). Representation as shared activity: situated cognition and Dewey's cartography of experience. *Journal of the Learning Sciences, 5,* 209–238.

Hall, R. (1999). The organization and development of discursive practices for "having a theory." In *Discourse Processes 27*(2) 187–218. Mahwah, NJ: Erlbaum.

Hall, R. (2000). Video recording as a theory. In A. Kelly & D. Lesh (Eds.), *Handbook of research design in mathematics and science education* (pp. 647–664). Mahwah, NJ: Erlbaum.

Hall, R., & Stevens, R. (1995). Making space: A comparison of mathematical work in school and professional design practices. In S. L. Star (Ed.), *The cultures of computing* (pp. 118–145). Oxford, England: Basil Blackwell.

Heath, S. B. (1983). *Ways with words: Language, life and work in communities and classrooms.* New York: Cambridge University Press.

Jones, G. (1989). Gender bias in classroom interactions. *Contemporary Education, 6,* 216–222.

Jordan, B., & Henderson, A. (1995). Interaction analysis: Foundation and practice. *The Journal of the Learning Sciences, 4,* 39–103.

Kaghan, W. N., & G. C. Bowker. (2001). Crossing boundaries and building bridges: Irreductionist "frameworks" for the study of sociotechnical systems. *Journal of Engineering and Technology Management, 18,* 253–269.

Kahle, J. B. (1990). Real students take chemistry and physics: Gender issues. In K. Tobin, J. B. Kahle, & B. J. Fraser (Eds.), *Windows into science classrooms* (pp. 92–134). London: Falmer Press.

Knorr-Cetina, K. (1981). *The manufacture of knowledge.* New York: Pergamon.

Kramarae, C., & Treichler, P. A. (1990). Power relationships in the classroom. In S. L. Gabriel & I. Smithson (Eds.), *Gender in the classroom* (pp. 41–59). Urbana: University of Illinois Press.

Labov, W. (1970). The logic of non-standard English. In F. Williams (Ed.), *Language and poverty.* Chicago: Markham.

LaFrance, M. (1991). School for scandal: Different educational experiences for males and females. *Gender and Education, 3,* 3–13.

Lakatos, I. (1976). *Proofs and refutations: The logic of mathematical discovery.* New York: Cambridge University Press.

Lakoff, R. (1975). *Language and women's place.* New York: Harper & Row.

Lampert, M. (1990). When the problem is not the question and the solution is not the answer: Mathematical knowing and teaching. *American Education Research Journal, 27*(1), 29–64.

Lampert, M. (1997). Teaching about thinking and thinking about teaching, revisited. In Virginia Richardson (Ed.), *Constructivist teacher education* (pp. 84–107). London: Falmer Press.

Lampert, M., Rittenhouse, P., & Crumbaugh, C. (1996). Agreeing to disagree: Developing sociable mathematical discourse. In D. R. Olson & N. Torrance (Eds.), *The handbook of education and human development: New models of learning and teaching schooling* (pp. 731–764). London: Basil Blackwell.

Latour, B. (1986). Visualization and cognition: Thinking with eyes and hands. *Knowledge and Society 6,* 1–40.

Latour, B. (1987). *Science in action: How to follow scientists and engineers through society.* Cambridge, MA: Harvard University Press.

Latour, B., & Woolgar, S. (1979). *Laboratory life: The construction of scientific facts.* Prineton, NJ: Princeton University Press.

Lave, J. (1988). *Cognition in practice.* Cambridge: Cambridge University Press.

Lie, M. (1995). Technology and masculinity: The case of the computer. *European Journal of Women's Studies. 3,* 379–394.

Linn, M., & Kessel, C. (1996). Success in mathematics: Increasing talent and gender diversity. In A. Schoenfeld, E. Dubinsky, & J. Kaput (Eds.), *Research in Collegiate Mathematics Education II* (pp. 101–144). Providence, RI: American Mathematical Society.

Lynch, W. T. (1994). Ideology and the sociology of scientific knowledge. *Social Studies of Science 24:* 197–227.

Martin, D. B. (2000). *Mathematics success and failure among African-American youth: The roles of sociohistorical context, community forces, school influence, and individual agency.* Mahwah, NJ: Erlbaum.

McCollum, P. (1989). Turn-allocation in lessons with North American and Puerto Rican students: A comparative study. *Anthropology and Education Quarterly, 20*(2), 133–156.

McDermott, R. (1993). The acquisition of a child by a learning disability. In S. Chaiklin & J. Lave (Eds.), *Understanding practice* (pp. 269–305). New York: Cambridge University Press.

McNeill, D. (1992). *Hand and mind: What gestures reveal about thought.* Chicago: University of Chicago Press.

Morse, L. W., & Handley, H. M. (1985). Listening to adolescents: Gender differences in science classroom interaction. In L. C. Wilkinson & C. B. Marrett (Eds.), *Gender influences in classroom interaction* (pp. 37–56). Orlando, FL: Academic Press.

National Council of Teachers of Mathematics. (2000). *Principles and standards for school mathematics.* Reston, VA: Author.

Noddings, N. (1992). Gender and curriculum. In P. W. Jackson (Ed.), *Handbook of research on curriculum* (pp. 659–684). New York: Macmillan.

Ochs, E. (1979). Transcription as theory. In E. Ochs & B. B. Schieffelin (Eds.), *Developmental pragmatics* (pp. 43–72). New York: Academic Press.

O'Connor, M. C., & Michaels, S. (1996). Shifting participant frameworks: Orchestrating thinking practices in group discussion. In D. Hicks (Ed.), *Discourse, learning, and schooling* (pp. 63–103). New York: Cambridge University Press.

Ong, M. T. (2002). *Against the current: Women of color succeeding in physics.* Unpublished doctoral dissertation. University of California, Berkeley.

Ortner, S. B. (1984). Theory in anthropology since the sixties. *Comparative Studies in Society and History 26,* 126–66.

Phelan, P. (1993). *Unmarked: The politics of performance.* London: Routledge.

Phillips, S. U. (1972). Participant structures and communicative competence: Warm Springs children in community and class. In C. B. Cazden, V. P. John, & D. Hymes (Eds.), *Functions of language in the classroom* (pp. 370–394). New York: Teachers College Press.

Phillips, S. U. (1983). *The invisible culture: Communication in classroom and community on the Warm Springs Indian Reservation.* New York: Longman.

Pollack, M. (2004). *Colormute: Race talk dilemmas in an American school.* Princeton, NJ: Princeton University Press.

Pomerantz, A. (1984). Agreeing and disagreeing with assessments: Some features of preferred/disprefered turn shapes. In J. Heritage & J. M. Atkinson (Eds.), *Structures of social action: Studies in conversation analysis* (pp. 57–101). Cambridge: Cambridge University Press.

Posner, T., & Horn, I. (1996). Unpublished manuscript. University of California, Berkeley, CA.

Richards, J. (1991). Mathematical discussions. In E. von Glasersfeld (Ed.), *Radical constructivism in mathematics education* (pp. 13–51). Dordrecht, Netherlands: Kluwer.

Roth, W. M., & McGinn, M. K. (1998) Inscriptions: toward a theory of representing as social practice. *Review of Educational Research, 68*(1) pp. 35–59.

Sacks, H., Schegloff, E. A., & Jefferson, G. (1974). A simplest systematics for the organization of turn-taking for conversation. *Language, 50,* 696–735.

Sadker, M., & Sadker, D. (1994). *Failing at fairness.* New York: Scribner.

Schegloff, E. A. (1992). On talk and its institutional occasions. In P. Drew & J. Heritage (Eds.), *Talk at work* (pp. 101–134). Cambridge: Cambridge University Press.

Schoenfeld, A. (1992). Learning to think mathematically: Problem solving, metacognition, and sense making in mathematics. In D. A. Grouws (Ed.), *Handbook of research in mathematics teaching and learning* (pp. 334–370). New York: Macmillan.

Schoenfeld, A. (2002). Making mathematics work for all children: Issues of standards, testing, and equity. *Educational Researcher, 31*(1), 13–25.

Shapin, S. (1989). The invisible technician. *American Scientist, 77,* 554–563

Spender D., & Sarah, E. (1980). *Learning to lose: Sexism and education.* London: The Women's Press.

Star, S. L. (1991a). Power, technologies and the phenomenology of standards: On being allergic to onions. In J. Law (Ed.), *A sociology of monsters: Power, technology and domination* (pp. 27–57). Sociological Review Monograph. No. 38. London: Routledge.

Star, S. L. (1991b). The sociology of the invisible: The primacy of work in the writing of Anselm Strauss. In David R. Maines (Ed.), *Social organization and social process: Essays in honor of Anselm Strauss* (pp. 265–283). New York: Aldine De Gruyter.

Swetz, F. J. (1994). *From five fingers to infinity: A journey through the history.* Chicago: Open Court Publishing.

Tannen, D. (1990). *You just don't understand.* New York: Morrow.

Tannen, D., & Bly, R. (1993). *Men and women: Talking together [Film].*

Tobias, S. (1990). *They're not dumb, they're different: Stalking the second tier.* Tucson, AZ: Research Corporation.

Tobin, K., & Garnett, P. (1987). Gender related differences in science activities. *Science Education, 71,* 91–103.

Traweek, S. (1988). *Beamtimes and lifetimes: The world of high energy physicists.* Cambridge, MA: Harvard University Press.

Volman, M., & ten Dam, G. (1998). Equal but different: Contradictions in the development of gender identity in the 1990s. *British Journal of Sociology of Education, 4,* 529–545.

Walkerdine, V. (1988). *The mastery of reason: Cognitive development and the production of rationality.* London: Routledge.

Willis, P. (1978). *Learning to labour.* London: Saxon House.

APPENDIX

Please see appended Table 4.4, which appears on pages 169–172.

Table 4.4

Uses of disagree and agree in January 19 lesson

Line number	Speakers	Utterance	With a person (P) or idea (I)	Disagree (D) or agree (A)
10	Sean:	Um, I—I—I just want to say something to Sheena, when sh— what she said about um that, that one, um—zero has to be an odd, an even number bec— I **disagree** because, um, because what what two things can you put together to make it?	Unclear	D
13	Sean:	Okay, um, I **disagree** with you because, um, if it was an even number, how—what two things could make it?	P	D
20	Teacher:	Um, there's a lot of **disagreement** about this issue, right? And you saw that the fourth graders who have been thinking about this for a long time also **disagree** about it, don't they?	I	D
			I	D
21	Mei:	Um, I h— I thought that zero was always going to be a even number, but from the meeting I sort of got mixed up because I heard other ideas I **agree** with and now I don't know which one I should agree with.	I	A
36	Teacher:	But he also said, I think, that all even numbers are made up of other even numbers.	Unclear	D
37	Mei:	I **disagree**.		
69	Cassandra:	I **disagree** with Sean when he says that six can be an odd number. I think six can't be an odd number because . . .look— [she gets up and comes up to the board]	P	D
137	Students:	I **disagree** . . .I don't think so . . .	Unclear	D
146	Teacher:	Do other people **agree** with that? Is that what you're saying, Sean?	I	A
148	Teacher:	Okay, do other people **agree** with him? [pause] Lin, you **disagree** with that?	P	D
149	Lin:	Yeah, I **disagree** with that because it's not according to like . . .here, can I show it on the board?	I	D
156	Sean:	. . .I **disagree** with myself . . .	P	D
172	Teacher:	Are people following this **disagreement**? This is an important thing that I didn't even realize we were **disagreeing** about, so it's important to see if we can try to figure this one out. What are you going to show?	I	D

Table 4.4 (*continued*)

Line number	Speakers	Utterance	With a person (P) or idea (I)	Disagree (D) or agree (A)
189	Teacher:	So you're saying the even numbers are the ones where you can group them all by twos and the odd ones are the ones where you end up with one left over?		
190	Lin:	Yeah, I think I **agree.**	I	A
219	Teacher:	Would somebody else like to come up and try a smaller number than twenty-one? Maybe one that everyone already **agrees** is odd? What's a number that everyone in the room already **agrees** is odd? Cassandra, what do you think is a number like that?	I	A
221	Teacher:	Does everyone in the room **agree** that seventeen is odd?	I	A
234	Tembe:	I **agree** and I have another way to do it.	Unclear	A
237	Teacher:	Again please?		
238	Tembe:	I **agree** and I have this other way to do it.	Unclear	A
251	Teacher:	Sean, what's your reaction to Ofala's definition?		
252	Sean:	Um, I'm not, I'm not—I **disagree** but, because um, I'm not, I'm not like using odd numbers to make them even and stuff, I'm using only even numbers to make them odd not, not. I'm not trying to make them but they can be odd, they can be odd . . . like ten and things like that. . .	Unclear	D
293	Ofala:	I don't **agree** that six can be odd.	I	D
298	Cassandra:	I **agree** with her. And I have a question for Sean. If six is odd, then what is five?	P	
311	Riba:	I **disagree.**	Unclear	D
314	Teacher:	Um hm. Cassandra, you can sit down. Lin, you can sit down. Thanks. Ofala, I'll just have you stand there for a minute since this is the definition you were suggesting, okay? Riba, what are you disagreeing with? Riba? What are you **disagreeing** with?	I	D
315	Riba:	With Sean	P	D

Table 4.4 (continued)

Line number	Speakers	Utterance	With a person (P) or idea (I)	Disagree (D) or agree (A)
320	Keith:	I **agree**.	P	A
321 322	Teacher: Keith:	You **agree** with Riba? Yeah	P	A
324	Ofala:	I **agree**. That's kind of like what I was going to say to him. Um, what Sean said doesn't have to do with my conjecture.	Unclear	A
325	Teacher:	Okay, so you're **disagreeing** that it's the number of groups. And Sean, are you saying that the number of groups matters?	I	D
332	Tembe:	I **disagree** with Sean.	P	D
337	Teacher:	You don't **agree** with that?	P	A
344	Sheena	I don't, I don't **agree** with that.		A
350	Ofala:	I **disagree** with um—	Unclear	D
351	Teacher:	Sean.	P	
352	Ofala:	. . . with Sean that um, you can't, you can turn an even, even number into an odd, because if you do, it will be like this [she uses her hand to cover one of the lines drawn on the board], they'll—it'll just change that whole number and make it into a, an odd.	P	D
387	Teacher:	Why would you like us to **agree** with you? Why would that be helpful if we all **agreed** that it was both odd and even? Can you convince us that that would be helpful?	P I	A
454	Bernadette:	Because um, if, if we were just trying to talk about one number, then you'd want to know if, if we, if we were just **agreeing** on even numbers or just **agreeing** on special numbers or just **agreeing** on even and odd. Or whatever.	I	A

Table 4.4 *(continued)*

Line number	Speakers	Utterance	With a person (P) or idea (I)	Disagree (D) or agree (A)
468	Teacher	Cassandra, do you **agree** or **disagree** with Jeannie?	P	AD
472	Teacher	Jeannie's saying it doesn't matter to her how many people, it matters if they change her mind. Are you **agreeing** with that or are you disagreeing with that?	I	AD
473	Cassandra	I **agree**.	Unclear	A
474	Teacher	What about you, Sean? A couple of times this week, you've had, you've taken a position that nobody else in the class **agreed** with. What does that make, does that make you change your mind, how does that make you feel?	I	A
475	Sean	It makes me feel fine.		
496	Teacher	He said his mother, when his mother said it, it was a little more convincing because he can trust her a little bit more? What do the rest of you think when you, when you're trying to figure stuff out, do you **agree** or **disagree** with Sean? Or what do you think about that? Maybe nobody has a comment about that. I don't know. Lin, what do you think?	P	AD
497	Lin	Um, I **agree** with Sean because you usually trust your mom because you live with your mom all those years. So you must—so you might trust your mom more than like the rest of the class because—or other people—because you haven't been with them. Maybe you've been with them as long you can remember, but not when you were just born like, your mom has been with you since you just born.	P	A
499	Sheena	I think was something um—um, I forgot what it was but this year I tried to get me to convince her about something that they—she didn't **agree** with me, she said that, that isn't it, so then finally I convinced her that it was it. . .	P	A
503	Ofala	Um, he **agreed**.	Unclear	A

Chapter 5

Commentary 1: Generating Useable Knowledge for Teaching

Elham Kazemi
University of Washington

ON GENERATING USEABLE KNOWLEDGE FOR TEACHING

Lewis begins this monograph with a statement about the fleeting nature of teaching; in her words, "its enactment flies by" (Introduction, p. 1). The same instructional moment cannot be recreated; each classroom conversation is different from the next. Given this natural predicament, the contributors to this monograph have pursued an enduring question: how can researchers portray classroom life in multiple ways so as to reveal the complexity of teaching to others?

In fact, the authors aim to contribute to efforts that make knowledge useable to teachers. Just what is meant for research to generate knowledge that is useable for teachers' everyday instructional practice? I found Lewis's elaboration helpful, and one that I draw on as I compare the chapters. Specifically, these chapters were brought together in the hopes of—

1. informing teachers as they engage in daily work in the classroom,

2. informing researchers on the complex work that is teaching, by expressing different analytic perspectives, and

3. creating opportunities for the reader to better understand those analytic perspectives themselves by comparing them.

After reading and rereading these chapters carefully, I considered how they viewed classroom life by comparing their analytic lenses. In making sense of the chapters, I turned to what Goodwin (1984) wrote about professional vision. He posits that professional vision consists of three practices:

1. Coding, which transforms phenomena observed in a specific setting into the objects of knowledge that animate the discourse of a profession

2. Highlighting, which makes specific phenomena in a complex perceptual field salient by marking them in some fashion

3. Producing and articulating material representations.

By applying these practices to scrutinized phenomena, participants build and contest professional vision, which consists of socially organized ways of seeing and understanding events that are answerable to the distinctive interests of a particular social group (p. 606).

Although individuals working within the same profession wrote the chapters in this volume, I found that Goodwin's conception of professional vision was useful in thinking about how the chapters generated different interpretations of the same instance of classroom practice. Each chapter—in its own way—created coding schemes, highlighting various aspects of the talk to allow us to see what was happening interactionally between the teacher and students in the same 6-minute episode. In so doing, they all tried to bring some coherence to the classroom events, preserved for us through videotape. From such coding and highlighting, they produced conceptual frameworks, inviting the reader to understand the sense they made of Ball's teaching moves and her students' participation.

The chapters each attempted to uncover underlying structures, principles, and interactional patterns that governed what happened. For Ball, Lewis, and Hoover (chapter 1), the governing factor is the structure of teaching that lies behind students' mathematical talk. For Schoenfeld (chapter 2), it is the principled way that a teacher makes decisions while engaged in the act of teaching. For Horn (chapter 3), it is a participant structure that allows for mathematically productive argumentation. For Posner (chapter 4), it is interactional patterns that raise concerns about equity and the opportunity to engage in authentic mathematical practices. I compare these frameworks to reveal how the collected analyses help us see more than any one on its own. I articulate where the analyses disagree, and consider whether the disagreements are resolvable. I end with questions that this monograph raises for our continued efforts to produce analyses that are useable for teaching.

COMPARING ANALYTIC VIEWS

Following Goodwin, "my own practices of seeing" are naturally part of my attempts to make sense of these chapters collectively. In an attempt to answer how the multiple views deepen our understandings at the same time that they reveal different things, I created a side-by-side chart on which I took notes on the analytic frame and claims each chapter made, line by line across the entire transcript. Using that representation, I made meta-comments about the frameworks that the chapters used to explain the line-by-line analysis. I then looked for convergence and divergence across the analyses and made my own interpretations about the effects of these similarities and differences on our understanding of the work of teaching and the work of research on teaching.

Two episodes within the 6 minutes emerged as common comparison points across the chapters. The first episode was the exchange between Sheena and Shea (S/S) at the very beginning of the transcript. The second was the extended exchange among Nathan, Betsy and others (N/B) about his statements regarding even numbers. Those two episodes allowed me to make a side-by-side comparison and helped me think about how these analyses made our understanding of the work of teaching more visible. In this commentary, I use those episodes to talk about the analytic perspectives themselves in how the authors coded, highlighted, and produced representations of what happened during the episodes.

Ball aims to help students develop mathematical knowledge through mathematical reasoning. She cultivates mathematical reasoning through the resources that become available as the teacher and children participate in public discussions. Those ideas reflect and express certain cognitive values. Goodnow (1990) writes that—

> cognitive development is marked by the acquisition of values. We do not simply learn to solve problems. We learn also what problems are considered worth solving, and what counts as an elegant rather than simply an acceptable solution. We do not simply acquire knowledge. We learn also that some particular pieces of knowledge are expected of us, that some can be happily ignored and that some are inappropriate for all but a few to own.... Finally, we learn that some ways of acquiring knowledge are more acceptable than others, and that particular ways of seeking are expected for particular areas of knowledge (pp. 259–260).

In Ball's vision of teaching, students must learn to engage with one another. Many may agree that making mathematical claims and producing mathematical arguments is an important aspect of mathematical practice. Central to Ball's teaching goals is that the "*public testing* [emphasis added] of ideas is crucial to doing and learning mathematics" (Ball et al., chapter 1, p. 31]). These analyses all hinge on this important dimension of Ball's teaching—that students are being encouraged to engage in mathematical argumentation *with one another.* Moreover, Ball aims to teach her students how to express their ideas, listen to one another, agree, disagree, and build on one another's ideas, all in a respectful manner. Ball and her colleagues underscore the fact that the three practices that make up their framework are in the service of drawing students into mathematical reasoning.

Other terms and phrases, used across the chapters, bring further contours to what is meant by engaging in argumentation. In Schoenfeld's (chapter 2) statements about Ball's knowledge and beliefs, we see such terms as *clarify, elaborate, extend, justify,* and *make clear arguments.* In Horn (chapter 3), such terms and phrases as *agree, disagree, justified position,* and *act on* or *defend a position* all highlight the cognitive values that the teacher aims to cultivate in this classroom.

Considering Schoenfeld's and Ball and colleagues' frameworks. In the Sean/ Sheena (S/S) exchange, Schoenfeld's (chapter 2) analysis helps us see how a predictable but flexible routine, initiated when the teacher asks the students to discuss a topic, underlies Ball's decision making. We see that a deep structure is evident in the way that students' claims are brought out for public consideration. Clearly, the routine has many decision points, which create many complex manifestations of how any particular discussion could unfold. Schoenfeld shows that the S/S exchange involves two cycles through the routine, and that the exchange with Nathan (lines 24–58) involves one cycle through the routine. According to Schoenfeld's modeling, how Ball navigates through these particular exchanges is understood through extensive consideration of the knowledge, beliefs, and goals that both buttress her teaching and become differentially activated during it. For example, when Sheena offered her first comment about the meeting, it warranted first being clarified and then being abstracted and reframed by Ball: "So you thought about something that came up in the meeting that you hadn't thought about before?" (line 7). The second cycle through the routine begins when Sean responded to the teacher's request for more comments about the meeting. This time, because Sheena asked right away for Sean to repeat what he said (thereby making the request for clarification), the teacher monitored the exchange between them because it fit within her goals and beliefs about what should be pursued. After listening and monitoring, Ball stepped in (line 20) to diffuse the tension created by Sean and Sheena's two ways of defining *even* to affirm that much disagreement still existed and that it was okay for Sean and Sheena to experience this disagreement. Although Schoenfeld's model is not a script, it fosters an understanding of the pathways Ball could use to navigate through any discussion.

Ball and her colleagues' (chapter 1) central concern is with how the teacher "read[ies] mathematical content so that students can engage in it ... and readies students to be doers and learners of mathematics" (chapter 1, p. 40). They posit that three related practices draw students into learning mathematical reasoning: naming, making claims, and evaluating mathematical claims. In both the S/S and N/B exchanges, they show us how participating in these practices plays out, paying specific attention to the way the teacher configures talk to engage students in those practices. I am thinking here, for example, how the analysis has us consider the way the teacher framed the question for discussion to convey the idea that doing mathematics is about exchanging and testing ideas. Ball asked her students to offer "comments" about the "meeting," and "listen to one another's comments, so that we can um, benefit from what other people say" (line 1). The phrasing of the request itself conveys to students the importance of hearing different ideas, readying them to engage in mathematical work.

Both Schoenfeld's and Ball and colleagues' analyses help us notice that the teacher works to clarify students' comments or prompt students to elaborate their thinking. Ball's responses to Sheena help her "understand and perform the work of 'making a comment'" (chapter 1, p. 24). Ball's prompting allows Sheena to specify the mathematical idea that she considered during the meeting with the fourth graders, and these more specific ideas, now made public, give more mathematical grist for the class to think about. Later, in the exchange with Nathan, Ball and coauthors' analysis again shows us that the teacher's questioning of Nathan is an example of Ball's putting a more formal mathematical idea about the sums of even numbers into language accessible to children, "when you put even numbers together, you get an even number" (line 26). Further, their attention to the practice of evaluating assertions allows us to see that in the N/B exchange, Betsy and others used counterexamples and raised questions about whether the claim works in general in the process of both clarifying Nathan's claim and evaluating its plausibility.

Listening to students' ideas brings with it a host of teaching questions. One set of questions centers on what teachers should do when students offer ideas. What should be pursued and when? Schoenfeld's analysis of why Ball did not intervene helps us consider that it is related to her goals of (a) allowing students to probe one another's ideas with respect and (b) encouraging students to think on their own. We also read in a number of chapters that Ball was hoping to engage the students in a meta-level conversation about the experience of having a meeting with another group to listen to others' ideas about mathematical issues that were perplexing them. All the chapters point out that one of the reasons the exchange between Sean and Sheena ends unresolved was that they are using different definitions of *even*. Because Ball has a complex set of goals, all of them cannot be highly activated at once. She decides to help diffuse some of the tension but attempts to return the students to thinking about the meeting.

We also know that Ball was thinking about having students consider, later in the lesson, a set of related conjectures about what happens when two even numbers are added. Ball and her colleagues and Schoenfeld offer that her listening and monitoring were also contributing to her *own* learning. The exchange they use to make this point is Ball's question to Nathan about his statements (line 26). I see this idea about teacher learning as being relevant to the S/S exchange as well because Ball could have been learning about what happens when students use different definitions during a mathematical argument. Whether Ball did or did not recognize the definitional dispute on the spot, the problem raises questions that are worthy of deliberation: how does the teacher help students reconcile and use different definitions in argumentation? How might definitional differences lead to disputes that are dead ends? When might such differences be generative? How might the teacher decide whether the class has the interactional resources and foundational knowledge to pursue particular disagreements? Studying those questions as they unfold in classroom discussions would add to the kind of useable knowledge that research could produce.

Layering on Horn's analysis. When I layered on Horn's (chapter 3) analysis of the S/S exchange, I saw the following relationship. Both Ball and colleagues' and Horn's frameworks focus on argumentation and how mathematical reasoning is supported in the classroom. Ball and coauthors identified practices—naming and using names, making claims, evaluating mathematical assertions—that pervade classroom talk. Horn's accountable argumentation could be thought of as a way these practices come together in an argumentation sequence when the practices of making claims and evaluating assertions lead to generative mathematical activity. Horn's analysis includes attention to the particular kinds of roles that students take on as they engage in argumentation. In the instance of the S/S exchange, the disagreement came to an impasse when the differing definitions and claims made by Sean and Sheena were not themselves examined and reconciled.

Although not discussed in detail in Horn's chapter, the N/B exchange (lines 45–58) is coded as another example of accountable argumentation; in that exchange, Betsy acts as the initiator of the argumentation sequence, and Nathan acts as the principal. Betsy's question and statements (line 48: You need three twos to make six. You can't put a four and a four or a …") serve to counter one version of the claim she thought Nathan made. This statement invites a response from Nathan to explain his idea further and from Betsy to acknowledge that she understands what he was trying to say: "So what you're doing is you're going by twos and then what two equals from then you go from—all the way up" (line 55). Betsy plays both the role of dissenter and that of clarifier in this short exchange, moving in fact from dissenter to ally once Nathan's statements are clarified. Nathan confirms that he was heard and says, "Yeah, I'm not going by every single number" (line 52). Ball and colleagues' claim is that students were able to engage in reasoning about claims publicly because they had learned to listen to, clarify, and evaluate one another's claims. The students take many turns without explicit teacher intervention, but again Ball and coauthors' chapter helps us see that the teacher was proactive in readying students for such work by focusing their attention on one another, restating claims for students to think about, and modeling ways of engaging with one another respectfully. Horn's description of accountable argumentation allows us to see that students must take on particular roles to accomplish such collective exchanges in which claims are evaluated. Melding Ball and colleagues' perspectives with Horn's, these roles, once recognized, might be explicitly cultivated by the teacher through talk as she readies students to engage in mathematics.

Bringing Posner's analysis into consideration. One idea that appears consistently in the chapters in this monograph is that agreeing and disagreeing publicly tend to bring to the fore a range of personal, social, and mathematical tensions. Ball and colleagues (chapter 1) state that the practices they define in their framework "provide ways to focus on who talks, what gets talked about, and how it gets talked about" (chapter 1, p. 41). This particular sentence resonated quite strongly for me when I read Posner's (chapter 4) analysis of the equity issues that are raised when studying the interactional patterns underlying classroom talk.

Posner's chapter, in this collection, is designated as the chapter on equity. We must take care, though, not to brand the other chapters as not concerned with equity. In fact, concern with equity—with how students are respected and heard in the classroom—is present in all the chapters. Ball and coauthors refer repeatedly to the importance of all students' being engaged in the practices they identify in their framework. Schoenfeld articulates one of Ball's central goals for the classroom community as being that all students feel "enfranchised in the classroom community, free to express themselves" (chapter 2, p. 96). This goal comes into focus when making sense of Ball's decision making. Horn too proposes her framework for accountable argumentation with a concern about how argumentation sequences unfold so that they are socially viable, without discernable negative consequences for participants. Through that lens, Ball and her colleagues show that students made many more claims than the teacher; and they, along with Horn, show that repeated sustained interactions occurred between students as they worked through mathematical issues. Yet what Posner helps the reader see is quite different. Through Posner's coding and highlighting, we have an opportunity to think specifically about how students perform and are designed into roles that render their contributions to be of lower status and perhaps less visible than others. Her analytic lens makes us think about how we know discussions are socially viable and whether all students are enfranchised.

Posner's analysis of the S/S exchange. Posner's (chapter 4) analysis of the S/S exchange has us consider the way Sheena and Sean positioned one another. As mentioned previously, all the chapters agree that this particular exchange ended with a dispute because Sheena and Sean were arguing on the basis of different definitions of *even.* Posner goes further to argue that the disagreement also revealed power differentials that are particularly alarming given that the exchange was between an African American girl and a white boy. Unlike the authors of the other chapters, she raises the possibility that Sheena's halting speech, her invocation of the fourth grader's arguments, and so on, all amount to conveying a reluctance to take a stand and be responsible for her position in the face of Sean's questioning. Sheena failed even to use her own understanding of the "two things make it" definition of *even,* which in fact was the current working definition for the entire class. And Posner has us consider whether Sean's move to engage Sheena with the statement "You said it was" (line 19) changed the nature of the disagreement from a conceptual to an interpersonal one. Posner uses additional data from the data corpus (beyond what is available in the 6-minute episode) to provide support for this position.

What enables Posner to see and make this argument? Her analytic views are guided by science studies, about how in the authentic practice of science, credit is given to individuals at the same time that it renders invisible others who contributed to a scientific development. She also draws on practice theory, which attends to the dialectic relationship between individual actions and social structures, thus coming to these transcripts with an eye to examining how everyday interactions produce and reproduce existing social structures that give differential power to groups in our society on the bases of race, class, and gender.

Posner's analysis of the N/B exchange. Posner's (chapter 4) analysis of the N/B exchange also departs markedly from the others. In the exchange with Nathan, she considers that part of the misunderstanding that was created when Nathan began talking might have resulted from confusing the everyday understanding of *even* as meaning "equal" with the property of numbers that the students were discussing. She makes this hypothesis after observing that Ball had noticed that confusion in his work. No one can say for certain whether Nathan was experiencing this confusion during the episode, but the larger point that teachers must develop keen insight into the relationship between students' everyday and mathematical understandings is important. Ball (1997) herself talks about the importance of attending to this idea, writing about how teachers' on-the-spot interpretations can miss what students are actually saying (we see evidence of this phenomenon in the transcript, in line 60–65, as well when Ball realizes that Sean has made a new conjecture only after Mei says she knows what Sean is talking about). A statement from a student that might seem confusing or completely disjointed may often have a certain logic behind it. One of the advantages of research is that it provides the time and opportunity for reflecting on classroom actions, which fly by at breakneck speed during class time. Such work can inform teachers in their efforts to better hear their students and consider the possibilities of how students' everyday understandings influence their emerging mathematical understandings.

Sheena's and Sean's contributions to class. One important limitation I see in Posner's (chapter 4) claims is whether Sheena's backing down in the moment, an African American girl to a white male, was related to a systematic pattern of backing down in the classroom in other circumstances, either by her or by others, along racial, class, or gender lines. From Ball and colleagues' chapter, we have a different view of what Ball does to support Sheena to make a statement that is publicly available for evaluation by the class. Ball and coauthors show the teacher prompting Sheena to make a more specific assertion about what she gained from the meeting (lines 2 through 7). They interpret Ball's moves as helping Sheena learn what it means to make a comment. As a reader, Posner's analysis of Sheena provoked other important questions for me about other students in the class. For example, what are the differences in how Mei and Sheena talked? Mei, it seems, did not back down. How should we understand Mei's positioning in the classroom and the factors that enabled her to display competence confidently?

Using data from Ball's teacher logs, Posner claims that Sean was also unsure of his ideas and confused definitions on more than one occasion. His idea about even and odd numbers ultimately got much play in the classroom, and the teacher's identification of his insight into a class of numbers called "Sean numbers" accorded it further status and play. I wonder how multiple perspectives on Sean, in relation to the class, would help us think about his contribution and the way it was highlighted in the class. Within race and gender, further class and linguistic differences can also be considered. In addition, knowing children's peer status and friendships would be greatly useful in this analysis, although this kind of data was not available to the authors here. To perceive Sean as the archetypical white male who gets all the attention would not be wise, nor would teachers' leaving this analysis thinking that all white males need to be curtailed in their contributions to class. The work on identity tells us that identities are fluid and changing, that children live and act across boundaries of many different groups and situations (Holland et al., 1998; Lave & Wenger, 1991; Sfard & Prusak, 2005). Would our analysis differ if we knew that Sean, for example, was the least popular child in the class or a child with high-functioning autism, and that the move to pay attention to his ideas made his thinking visible and allowed him to display competence in a way that improved his social status or made him better able to consider other people's point of view? Would our perception differ if we saw changes over time in Sheena's ability to articulate her position and not back down when challenged? Do children in this class engage differently when they are newcomers to the school versus old-timers?

The consequences of naming "Sean's numbers." The practice of naming ideas after particular students is something I have seen many teachers use (and which I

use as well). As a strategy is shared, a teacher or even children might start referring to it by using a classmate's name, "Darlene's arrow method," for example. If
thought about in relation to the norm of creating a base of public knowledge in the
classroom, such practices can be quite useful because they enable the class to refer
back to particular ideas; once named and defined, ideas can be carried across to
other lessons and be connected to one another. Posner's (chapter 4) analysis of the
events that led Ball to label all numbers having the property that Sean first noticed
about six—a property that every other even number has—as "Sean numbers" calls
this practice into question. Posner suggests that it is not a neutral activity but may
have rendered invisible the contributions of Mei and Riba in producing the generalization. We see the class refer to other ideas by similar labels: Sheena's working definition of *even,* Ofala's definition of *odd,* for example. Should these kinds
of labels always be questioned? Sheena's definition, we learn, was also modified
by Mei, who often contributed to class discussions (something we learn across
Ball's other papers; see, for example, Ball, 1997). But Posner does not specifically
suggest that it too should have been named for Mei and Sheena. Naming, in Ball
and colleagues' (chapter 1) framework, is an important dimension of mathematical practice, and thus a crucial consideration for future research and discussion
among educators and researchers is how the generation of class-derived labels
(named after particular students) reflects messages about who is competent in
mathematics. Posner's critique of the practice of naming ideas after students opens
up much for us to consider.

The role of equity across analytic views. My attempts to make sense of the four
analytic frames led to me other questions about the role equity plays in the analytic lenses used across the four chapters. Should every analytic framework consider who participates and how, or is the lesson from these analyses that the ability
to layer analytic views is of utmost importance in allowing us to see teaching and
produce useable knowledge? What I am asking is whether Ball and colleagues'
framework should or could be expanded to pay particular attention to who, in what
way, and how often students and the teacher make and evaluate claims? Could
Schoenfeld's model include in the teacher's decision making more explicit consideration of what the teacher knows about students' ways of engaging and concerns
about students' status and position? Might Horn's framework attend to whether
patterns are evident in the kinds of roles students play in classroom discussions
and whether those roles contribute to status differences?

SUMMARY AND FUTURE CONSIDERATIONS

As one way of summarizing my analysis of these chapters' contributions to a
multilayered view of teaching, let me state the considerations about teaching that
the chapters invited us to see. These chapters have contributed to the following
complex view of teaching and the resources that became available for learning as
Ball aimed to engage her students in mathematical argumentation:

1. Ball and colleagues' (chapter 1) practices of naming, making claims, and evaluating assertions help us see the mathematical activity that underlies the teacher's and students' public contributions to discussions.

2. Schoenfeld's (chapter 2) routine conveys how ideas enter into public consideration. The routine for discussing a topic makes Ball's decision making visible and predictable rather than ephemeral and unattainable as acts of a gifted teacher.

3. Horn's (chapter 3) accountable argumentation structure allows us to consider how mathematical reasoning is supported by the roles teachers and students play when public disagreement arises over a mathematical assertion. The discourse structure is a way to think about the trajectory of a discussion, whether and how it is both socially viable and mathematically productive.

4. Posner's (chapter 4) attention to equity issues leads us to further consider the impact of students' positioning in discussions—how students' contributions are heard, considered, and affirmed by the class and by the teacher.

Each of these major analytic views is directly related to the work of teaching, when the aim is to make mathematical argumentation a collective enterprise. One productive avenue for further collaboration and analysis is to explore whether these analytic views can be considered together—can researchers themselves model how to do so? What additional work might we do to create a unifying framework that allows us to consider these major views? What would constitute such a framework? Perhaps, like Rogoff's (2003) theory of human development, we may find that the frameworks can be brought together by thinking about the various planes of analysis and what comes into focus within each plane. Whatever the product, the field of teaching and learning would be advanced by such synthetic work.

This monograph also inspires much other research related to the cultivation of classroom communities. The chapters point to the fact that by January the teacher has done much work that enables the students to engage in argumentation. Teachers and researchers will surely want to know the kind of work teachers have to do to make that environment possible. Creating discussion-intensive classrooms in which argumentation is fostered produces its own inherent complexities. How do students learn to agree and disagree? How much do they contribute, and how are ideas taken up by the group? What are the social relations among children? Posner emphasizes race and gender in her discussion of equity issues that undergird the talk in the 6-minute segment. Issues of class and linguistic competence should be considered. If one thinks about accountable argumentation as a particular structure—and agrees that it could be mathematically generative—then how does one think about setting it up? When does it become socially viable, and when not? How can teachers be attuned to the way students' participation in classrooms reproduces broader social inequities that divide us along racial, class, and gender lines? How does one think about whether something is mathematically productive? A consideration of those factors would fit with Schoenfeld's idea that his decision-making model can become a source of serious consideration among teachers—what decisions could they have made, and why?

Although understandably limited by the data set, students' own experiences of these events, nevertheless, are absent. Many claims are made in these chapters about what students are learning and what kinds of selves are being developed. The students' own perspectives are missing. What is being a newcomer in this classroom like? What is developing competence in English like in such a discussion-intensive classroom? What happens to students who do not like to talk publicly in such classrooms? What happens to friendship patterns when students agree and disagree with one another? As we continue to investigate the teacher's role in classroom discussions, we must include students' own interpretation of classroom events. Such data will surely complicate our own data collection and analytic procedures, but these complications would also enrich our understandings.

As I read and thought about the chapters in this monograph I was reminded again and again of Paley's (1986) words in her essay on the importance of listening to children's ideas:

> It is curiosity, not answers, that we model. As we seek to learn more about a child, we demonstrate the acts of observing, listening, questioning, and wondering. When we are curious about a child's words and our responses to those words, the child feels respected. The child *is* respected. "What are these ideas I have that are so interesting to the teacher? I must be somebody with good ideas" (p. 127).

The chapters in this monograph show deep respect for teachers, teaching, and children. They provoke a lot of serious thinking about what children and teachers do, what kinds of understanding and values are expressed and cultivated through teaching. And they demonstrate a deep commitment to taking the work of children and teachers seriously. Coming together to focus on a tiny slice of life in a classroom, these chapters show collectively how much we can learn about teaching as they point simultaneously to the work we still must do.

REFERENCES

Ball, D. L. (1997). What do students know? Facing challenges of distance, context, and desire in trying to hear children. In B. J. Biddle, T. L. Good, & I. Goodson (Eds.), *International handbook of teachers and teaching* (pp. 769–818). Dordrecht, Netherlands: Kluwer.

Ball, D., L., Lewis, J., & Hoover, M. (2007). Making mathematics work in school. In Alan H. Schoenfeld (Ed.), *A study of teaching: Multiple lenses, Multiple views, Monograph no. 14 of the Journal for Research in Mathematics Education* (pp. 13–44). Reston, VA: National Council of Teachers of Mathematics.

Goodnow, J. (1990). The socialization of cognition: What's involved? In J. W. Stigler, R. A. Shweder, & G. Herdt (Eds.), *Cultural psychology: Essays on comparative human development* (pp. 259–286). New York: Cambridge University Press.

Goodwin, C. (1984). Professional vision. *American Anthropologist, 96*(3), 606–633.

Holland, D., Lachiotte, W., Skinner, D., & Cain, C. (1998). *Identity and agency in cultural worlds.* Cambridge, MA: Harvard University Press.

Horn, I. (2007). Accountable argumentation as a participation structure to support learning through disagreement. In Alan H. Schoenfeld (Ed.), *A study of teaching: Multiple lenses, multiple views,* Monograph no. 14 of the *Journal for Research in Mathematics Education* (pp. 97–126). Reston, VA: National Council of Teachers of Mathematics.

Lave, J., & Wenger, E. (1991). *Situated learning: Legitimate peripheral participation.* Cambridge, England: Cambridge University Press.

Lewis, J. (2007). "Through the looking glass": A study of teaching. In Alan H. Schoenfeld (Ed.), *A study of teaching: Multiple lenses, multiple views,* Monograph no. 14 of the *Journal for Research in Mathematics Education* (pp. 1–12). Reston, VA: National Council of Teachers of Mathematics.

Paley, V. G. (1986). On listening to what the children say. *Harvard Educational Review, 56,* 122–131.

Posner, T. (2007). Looking at equity in mathematics classrooms. In Alan H. Schoenfeld (Ed.), *A study of teaching: Multiple lenses, multiple views,* Monograph no. 14 of the *Journal for Research in Mathematics Education* (pp. 127–172). Reston, VA: National Council of Teachers of Mathematics.

Rogoff, B. (2003).*The cultural nature of human development.* New York: Oxford University Press.

Schoenfeld, A. H. (2007). On modeling teachers' in-the-moment decision making. In Alan H. Schoenfeld (Ed.), *A study of teaching: Multiple lenses, multiple views,* Monograph no. 14 of the *Journal for Research in Mathematics Education* (pp. 46–96). Reston, VA: National Council of Teachers of Mathematics.

Sfard, A., & Prusak, A. (2005). Telling identities: In search of an analytic tool for investigating learning as a culturally shaped activity. *Educational Researcher, 34,* 14–22.

Chapter 6

Commentary 2: Moving From Shared Data to Shared Frameworks

Miriam Gamoran Sherin
Bruce L. Sherin
Northwestern University

Too often in educational research, the standards for what counts as scientific consensus are lamentably poor. We agree that "constructivism" is a good idea, that learning should be "authentic," and that teachers' "pedagogical content knowledge" is important. But rarely are we precise about what our beliefs mean, nor about the data that, as a community, we believe stands as solid evidence for those shared beliefs.

We suggest that the work presented in this volume holds the promise of taking the field a step forward. Specifically, an examination of a common data set by a group of researchers appears to be a useful context for considering how we might move toward a well-articulated and well-supported consensus among the researchers involved. Although sharing data is certainly not a necessary condition for forging consensus, we believe that much is to be learned by considering how different accounts of the same short video excerpt might be unified.

To be clear, the authors of this volume did not set out to produce a shared theory of teaching. In fact, their stated purpose is quite the opposite. In her introduction to this monograph, Lewis explains that "we intentionally seek alternative and competing perspectives on problems of practice because no one theory will sufficiently illuminate what is by nature a complex object of study" (Introduction, p. 5). In the long run, Lewis may be proved to be correct; no one theory may turn out to be sufficient to explain classroom practices. We nonetheless believe that the "alternative perspectives" orientation is weaker than necessary. Instead, we argue that an attempt to forge consensus around a common set of theories or models is crucial, no matter how difficult that task might appear to be at present.

PROBLEMS AND POTENTIAL

To begin, we want to briefly lay out why a volume of this sort calls out for a synthesis based on shared models and theories. First, despite different purposes and analytic approaches, many commonalities are evident across the analyses—all examine, albeit to different degrees, argumentation, justification, and a notable interaction between Sean and Sheena. But although the various chapters in this monograph look at the same episode, and even attend to similar phenomena, whether they are in agreement is difficult to tell. For example, are the elements of mathematical talk identified by Ball, Lewis, and Hoover (chapter 1) consistent with the description of accountable argumentation given by Horn (chapter 3)? Are they complementary? And what is the relationship between Horn's discussion of discourse as shifting from accountable argumentation to peer dispute and Posner's (chapter 4) characterization of discourse as moving from being conceptually based to being interpersonal (Engle & Greeno, 1994)? Similarly, we wonder about explicit and implicit connections between Ball and her colleagues' description of the processes involved in evaluating claims and Schoenfeld's (chapter 2) discussion of the multifaceted routine through which the teacher responds to students' mathematical ideas.

The need for an encompassing framework can also be seen when looking closely at the individual articles in this volume. Consider, for example, the chapter by Ball and others (chapter 1). In discussing the particular practices that are the focus of their chapter, the authors state that their analysis led them "to posit three essential elements that undergird the nature of mathematical talk in the segment." Their list of elements clearly has intuitive appeal—the students' talk does seem to be atypical in the ways that mathematical terms are used, in the students' offering of claims, and in the persistent probe for justifications of claims offered. Not clear, however, is the extent to which Ball and her colleagues are suggesting that those elements comprise the entire landscape of classroom discourse in the focus episode. Are those elements *the* three elements? Do other foundational elements of the classroom talk exist? And if so, how would they be identified? In other words, how seriously are we supposed to take the particular decomposition of mathematical talk embodied in this list? Do two different kinds of naming, or two different kinds of mathematical assertions, actually exist?

Our point, of course, is that none of those questions can be answered absent a broader framework in which all the theoretical entities are embedded. That is, if those entities are *elements,* being so implies that they should somehow be components of something larger. And if they are components, being so suggests a particular kind of decomposition, according to some overarching logic. What is that logic?

Our intent is not, in particular, to be critical of the contribution by Ball and her colleagues (chapter 1). Indeed, in our view their work is representative of the state of the art in our field. Our point, instead, is to attempt to make clear what is possible if researchers come together with the sort of shared focus attempted in this monograph. Real progress will not be achieved, we believe, only—or even primarily—by looking at the same data. Instead, it will be achieved when researchers forge consensus on frameworks and models.

In what follows, we examine the extent to which the chapters in this monograph foster insights about the nature of such consensus. Specifically, we describe two approaches to synthesizing the work of the researchers represented here. First, however, we briefly introduce each of the chapters in the monograph. Our intention in doing so is not to summarize the main points of each chapter, but rather to point to particular perspectives that will be relevant for the syntheses that we present later on.

THE INDIVIDUAL CHAPTERS

In Lewis's introduction to the monograph, she emphasizes the complex nature of teaching. It is an activity filled with uncertainties, in which the same lesson plan is likely to play out differently in multiple enactments. As Lewis explains, this uncertainty poses challenges for teachers as well as for those who wish to study teaching. How can we make sense of such a complicated endeavor? What would constitute understanding an episode of mathematics teaching? The four remaining chapters in the monograph examine this issue by considering the same classroom episode, the first 6 minutes in a lesson that has come to be known as "Sean numbers" (Ball, 1990, 1993).

Ball and her colleagues (chapter 1) investigate the mathematical work in which students are engaged and the role of the teacher in establishing and supporting that work. The task is challenging, particularly because the teacher in the episode, Ball herself, can seem "invisible" at first glance. In contrast with the central position often held by teachers during whole-class discussion, in this episode, a great deal of student-to-student talk takes place without interjections from the teacher. Furthermore, the teacher's stated agenda is to hear from students; she asks them for the comments they have about the previous day's meeting with the fourth graders. Thus, the specific mathematical focus of the lesson is dictated, in large part, by the students (Hufferd-Ackles, Fuson, & Sherin, 2004).

In examining the classroom talk in this segment, Ball and coauthors (chapter 1) focus on three elements: (a) "naming," that is, the use of words and phrases to describe mathematical content and the practice of learning mathematics; (b) an orientation to making claims about mathematics and about doing mathematics; and (c) the evaluation of mathematical assertions. The authors propose that each element supports the work of doing mathematics, and they provide detailed evidence of students' participation in those discourse practices during the 6-minute episode under consideration. In addition, a central goal of Ball and colleagues' chapter is to make visible how the teacher's own discourse is used to establish and reinforce those discourse patterns. Thus, we understand Mei's reply, "I'm going to listen more to the discussion and find out," not just as a response from an extraordinary third grader but as a response to her teacher's request to "listen to one another's comments, so that we can benefit from what other people say." Ball has explicitly named the activity of listening and elevated it to a position of significance in supporting one's own learning of mathematics.

The second chapter concerns Schoenfeld's model of the teaching process. His approach involves an attempt to describe, at the moment-by-moment level, the decision-making process of a teacher. In prior work, Schoenfeld (1998, 1999) proposed models of two secondary level teachers, both of whom maintained a strong presence in whole-class discussion. Attempting to model the teaching of Deborah Ball in this episode was therefore an important test case for examining the limits of the model.

Schoenfeld's (chapter 2) model encompasses a teacher's goals, beliefs, knowledge, and actions. Because a teacher may hold multiple goals simultaneously, the model allows for the shifting of goals over time and for different sets of goals to be given priority at particular moments. To model Ball's teaching in the episode under investigation, Schoenfeld identifies a "flexible, interruptable routine," in which comments made by students are considered in light of Ball's current goals, thereby resulting in corresponding actions on the part of the teacher. Ball cycles through the routine five times in the 6-minute episode. And although her responses to students appear spontaneous and context-dependent, Schoenfeld's model suggests that they are not arbitrary. In contrast, Schoenfeld argues that Ball's responses in this episode can, for the most part, be predicted on the basis of her goals, beliefs, and knowledge. For example, the model draws on particular beliefs about students and about mathematics to account for why Ball does not intervene in Sean and Sheena's conversation, but yet soon after, takes a detour from her stated purpose to explore Nathan's ideas about the composition of even numbers.

The third chapter, by Horn, examines the following teaching dilemma: How can classroom discourse promote productive discussion of competing mathematical claims (Ball, 1996; Sherin, 2002; Silver & Smith, 1996; Wood, 1999)? To investigate this issue, Horn introduces a discourse structure she calls "accountable argumentation." According to Horn, accountable argumentation "organizes the public disagreements among students" (chapter 3, p. 104) in such a way that the mathematical focus of discussion is maintained and social discomfort is minimized. Horn presents warrants for classifying a disagreement as an instance of accountable argumentation, noting specific norms and expectations, roles for participants, and the use of historical information as distinguishing elements. Furthermore, she illustrates that mathematical learning can and does take place during such disagreements. As an example, Horn deconstructs Sean and Mei's discussion about the parity of 10. In an unforgettable moment from the "Sean numbers" discussion, Mei responds to Sean's claim that "six can be an odd and an even number" by asking, "Why do you not call ten [an] … odd number and an even number?" Horn uses this example to illustrate that accountable argumentation can be both sustained by students and mathematically rigorous. Toward that end, she highlights that this conversation moves from the particulars of six to more generalized claims about the oddly even numbers. (A discussion of the oddly even and evenly even numbers can be found in Posner's contribution to this volume.)

In the fourth chapter, Posner explores the social nature of mathematics learning in the classroom episode, with a particular focus on issues of equity. Posner looks closely at two segments and attempts to explicate both the mathematical meaning of students' comments as well as the interactional meaning of students' statements and actions in those segments. She explains that doing so "can provide a lens to view the roles, social relationships, and power relationships among participants" (chapter 4, p. 136).

For each segment, Posner (chapter 4) brings a variety of lenses to her analysis. For example, she considers (a) the nature of the mathematical ideas raised by students in the discussion and how those ideas have been treated by the class previously, (b) the history of students' participation in the classroom, (c) the use of inscriptions and the positioning of different inscriptions in this classroom, and (d) the degree to which different students are willing to assume authoritative roles in this discussion. Thus, to explore Sean and Sheena's discussion about the parity of zero, Posner suggests what each understands about even and odd numbers, the kind of authority they turn to for justification of those ideas, how much they typically talk in class, and more. We are to understand, for instance, that although Sheena "seemed to have a clear grasp of the nature of even and odd numbers," she tended to attribute "the justification of her positions to third parties (mainly boys)" (p. 145). This tendency on Sheena's part raises questions, then, about her comment about zero, "I'm not saying that it has to be even. I meant that it could be," (p. 143). As Posner asks, "Did Sheena change her understanding about the nature of zero through the work of the disagreement?" (p. 143). Or did she back off to avoid a confrontation with Sean? Posner's analysis clearly portrays the complexity of classroom interactions as well as the complexity involved in interpreting such interactions.

TOWARD A SYNTHESIS

What, then, might constitute a synthesis of these diverse approaches? And what would a grand synthesis tell us about the nature of teaching? We are limited—by both space and our own ability—in what we can do in this commentary. As a start, however, we illustrate two types of syntheses that we believe are useful. Both approaches shed light on the unique contributions of the chapters in this volume while suggesting important next steps for the future.

Our first approach consists of an ontological synthesis of the articles in this volume. In other words, we look closely at the kind of theoretical entities that these researchers make reference to as they make sense of the teaching episode. In doing so, we ask how those elements might fit together in a broader framework. Because all four chapters use classroom discourse as a central lens through which to examine the teaching episode, we have chosen discourse as the starting point for this synthesis. To be clear, albeit having a common focus, the chapters explore different components of the discourse that takes place, for different purposes.

Consider the work of Horn (chapter 3), for example. In making claims that accountable argumentation is a discourse structure, Horn is in fact making claims about what constitutes a classroom discourse structure more broadly. In particular, she suggests that disagreements can be classified by norms, expectations, and interactional roles. Moreover, Horn identifies a specific set of expectations that distinguish instances of accountable argumentation from other forms of whole-class discussion. Similarly, she presents the range of roles that participants take on during accountable argumentation. In these ways, Horn offers a detailed character-ization of accountable argumentation and also identifies what she sees as essential dimensions of participant structures during whole-class discussion. Recall, also, that Horn gives an example in which an instance of accountable argumentation is transformed into a "peer dispute." Thus, Horn implicitly states the existence of multiple kinds of discourse structures that one might find.

Ball and her colleagues (chapter 1) take a different approach to identifying the substance of classroom discourse during the episode under consideration. We be-lieve that the three elements that are the foci of Ball and her colleagues' analysis are not discourse structures in the sense that we have discussed previously. Rather, we believe that they are best thought of as constituent elements of a range of dis-course structures. For example, "naming" might take place within an instance of accountable argumentation as well as during a peer dispute. In fact, juxtaposing Ball and coauthors' and Horn's analyses, we find explicit evidence of "making claims" (the second of Ball et al.'s discourse elements) across both of those activi-ties. Specifically, in the accountable argumentation portion of Sean and Sheena's conversation, Sheena states, "I could show you it" and proceeds to use the number line to demonstrate the reasoning behind her claim that zero is an even number. Later, as their conversation moves into a peer dispute, we find Sheena continuing to make claims, as in her statement "But that doesn't mean it always is even." We suspect, in fact, that the discourse elements described by Ball and her colleagues might be found across a broad range of discourse structures.

Some other ontological features of Ball and coauthors' (chapter 1) analysis de-serve mention here. In their analysis, Ball and her colleagues have chosen to rei-fy—to treat as first-class theoretical entities—discourse *processes,* such as nam-ing. But their analysis is populated, at least implicitly, with a variety of other kinds of entities; for example, with the process of naming are the names themselves. Also present are the particular *assertions* that students make.

Decisions about what we choose to treat as first-class theoretical entities will be significant as we attempt to move toward consensus. Ball and colleagues' (chapter 1) decision to foreground the processes rather than some of the entities that par-ticipate in those processes is potentially crucial. On the one hand, the foreground-ing of processes might be precisely the appropriate move if we want to get a theo-retical handle on complex classroom events. On the other hand, the reification of processes may contribute to an ontological vagueness that causes more difficulty than necessary in comparing analyses across research projects.

Posner (chapter 4) can be seen, at times, as using approaches similar to those of both Horn (chapter 3) and Ball and colleagues (chapter 1) to make sense of the classroom discourse. In particular, like Horn, she describes discourse structures; she explicitly contrasts conceptually based disagreements with interpersonal dis-agreements (Engle & Greeno, 1994). Rather than cite specific warrants that dis-tinguish those two types of conversations, however, Posner relies on broad charac-terizations of differences in participants' position, motivation, and understanding of mathematics and of classroom norms.

Later in her article, Posner (chapter 4) focuses on the act of attributing names to mathematical ideas that arise in the classroom. Thus, the oddly even numbers become referred to as "Sean numbers," and the definition of even numbers as objects in which "two things make it ... without using halves" comes to be known as "Sheena's definition." This idea seems clearly related to Ball and others' (chapter 1) discussion of "naming" as a process that directs students' attention to specific objects, tools, and ways of learning and doing mathematics. The crucial aspect for this synthesis, however, is that both can be understood as constituent elements of the classroom discourse and are likely to be found within a range of discourse structures.

A SECOND ATTEMPT AT SYNTHESIS

Schoenfeld (chapter 2) takes an approach that is quite different from those in the other chapters in this volume. Rather than attempt to categorize and deconstruct kinds of discourse, Schoenfeld focuses on elucidating some of the mechanisms that *generate* particular patterns of interaction. His goal is to describe, at the level of mechanism, what a teacher does and why. In our search for a grand theory, we found that Schoenfeld's chapter contains, most explicitly, a *model* of a phenomenon. Because his focus and approach are somewhat different from those of the other articles, we discuss what would be involved in forging a synthesis in which we start from Schoenfeld's model.

Despite the fact that Schoenfeld's (chapter 2) model addresses teaching at a broad level, it would nonetheless need to be significantly expanded if it is to address the types of insights contained in the other articles in this monograph. To begin, note that Schoenfeld's model constitutes a strongly teacher-centered analysis of classroom events. It clearly sees the unfolding of classroom events as being largely determined by factors outside the teacher. In Schoenfeld's model, those factors are modeled as inputs to the flowchart at various points. Thus, Schoenfeld's model is consistent with the observation that extra-teacher factors are involved, but those factors are not themselves modeled. That omission is not necessarily a problem with Schoenfeld's approach; when we model, we must narrow our focus and simplify. But if we want to make contact with the other analyses in this volume, we need to push Schoenfeld's model outward to encompass those extra-teacher factors.

Specifically, the other three articles in this volume all take, as their unit of analysis, interacting units that are larger than the teacher. For example, Ball and her colleagues (chapter 1) consider how students respond in light of discourse norms established previously by the teacher. This focus is, in a sense, the reverse of Schoenfeld's approach to looking at how the teacher responds to students. Moreover, Horn (chapter 3) and Posner (chapter 4) look closely at how students respond to one another. That kind of interaction is not captured by Schoenfeld's current model. One productive place to consider expanding might be in the practice of evaluating claims. Ball and her colleagues present a three-phase structure involved in the evaluation of claims—moving from clarifying a claim, to providing examples to support or refute a claim, and finally, to considering whether a claim is true in general. One way to expand Schoenfeld's model to encompass those insights might be to enrich the teacher model so that it is more explicit about some of the classroom contingencies on which the teacher's decisions are based. More elaborately, we could attempt to build a model of student behavior in those classroom discussions, patterned after the model for teachers, that interacts with the teacher model. The resulting interaction among a "teacher model" and a set of "student models" raises questions, however. For example, is the whole truly greater than the sum of its parts—would a characterization of each individual voice in the classroom capture the workings of the entire class?

Schoenfeld's (chapter 2) model also needs to be expanded in another respect as we attempt to create a synthesis across the chapters in this volume. So far, we have noted that we can expand our unit of analysis to more thoroughly model the unfolding of classroom events. But we can also imagine including, in our modeling efforts, more of the larger history of a classroom— the sort of unfolding that happens over days, weeks, and months rather than over the minutes of individual classroom events. That sort of expansion is likely needed if we want to begin to encompass the sort of phenomena that are central in the article by Posner (chapter 4).

Certainly, the nature of the endeavor we have mapped out here is potentially mammoth; it is equivalent to the full task of understanding teaching and learning. But if we want to fully reap the benefits that can be extracted from multiple analyses of the same classroom episode, then we believe we must at least attempt that sort of synthesis.

THE VALUE OF SHARED VIDEO

In the preceding sections of this commentary, we called into question the assumption that the study of teaching necessarily requires work from different perspectives. In particular, we argued that sharing models and theoretical frameworks is a crucial goal that should underpin any attempts at analysis of common data. However, this view does not mean that we believe that sharing video data is not useful; on the contrary, we believe that sharing video data can advance the work of our community in a number of important ways.

First, sharing our video data may be helpful even if other researchers do not pursue their own competing analyses. In most studies, other researchers have access to our "raw" data only through the transcripts that we embed in our published articles. But having access to the raw data at some level is absolutely necessary for other researchers to truly evaluate the claims that we make. To date, only a few studies have published video data alongside written research reports (Carraher & Nemirovsky, 2005; Sfard & McClain, 2002). This monograph is certainly noteworthy in that respect.

Second, shared video episodes are also valuable for the field because they provide common reference points for discussion, even if researchers do not provide full-fledged competing analyses of the video episodes. They can become part of the shared vocabulary of our disciplines, and they can become shared touchstones of the ideas we generate. Thus, aside from the potential of shared video data as the focus of research, we have benefited from being able to call on the widely known "Sean numbers" video in discussions with colleagues.

Third, at the highest end, we might want to work toward the collection of true shared video libraries that are the focus of analysis by many researchers. Certainly, quantitative research has benefited from the existence of shared national databases. But the creation of a shared video archive would present numerous difficulties. Matters of privacy and the protection of human subjects would need to be carefully considered. And whether such a video archive would prove useful is far from clear. The very richness of video can present difficulties in using the data without a great deal of supporting knowledge of the context in which those data were collected. In our own research, we have noted that an analysis of video data can be difficult when we have not been present at the time the data were collected.

Hiebert, Gallimore, and Stigler (2003) describe an instance in which they were using videos from the 1999 TIMSS project with a group of mathematics teachers at a professional development event. One of the teachers in the audience happened to have participated in the TIMSS project and had agreed to have her videos shared publicly. When the teacher was introduced to the group, the participants spontaneously applauded. As Hiebert and colleagues explain, "[T]hose assembled were not applauding the lesson Ms. Lancour had taught. They had not seen her lesson. They were applauding her courage in allowing others to view the lesson as a means of improving their own mathematics teaching" (p.56). Hiebert and his colleagues refer to that teacher as one of the "new heroes of teaching" for her willingness to share her teaching in that way. We similarly applaud Deborah Ball and thank her for the opportunity not only to enter her classroom but to study it.

We remember vividly our first viewing of the "Sean numbers" video at the NCTM conference in April, 1991. The room was packed with teachers, teacher educators, and researchers. At one point Ball asked us to focus on Ofala, and yet Miriam could not get her mind off Mei's comment that "if all numbers were odd and even, we wouldn't be even having this discussion" while Bruce kept thinking of questions *he* wanted to ask Sean and Sheena. We have viewed the video many times since then, at conferences and meetings, and, with Ball's permission, with teacher education students at Northwestern University. And each time, we continue to be captivated. The video prompts researchers, policy makers, and teachers to recognize the depth at which students can engage with mathematics and to want to understand how this outcome is possible. The chapters in this volume provide important perspectives on that question. By pushing forward from common data to common frameworks, we believe the field can move even closer to understanding the nature of teaching.

REFERENCES

Ball, D. L. (1990). *Sean numbers: Transcript from Deborah Ball's classroom, January 19, 1990.* East Lansing, MI: M.A.T.H. Project, College of Education, Michigan State University. Used with permission.

Ball, D. L. (1993). With an eye on the mathematical horizon: Dilemmas of teaching elementary school mathematics. *Elementary School Journal, 93*(4), 373–397.

Ball, D. L. (1996). Teacher learning and the mathematics reforms: What we think we know and what we need to learn. *Phi Delta Kappan, 77,* 500–508.

Ball, D. L., Lewis, J., & Hoover, M.. (2007). Making mathematics work in school. In Alan H. Schoenfeld (Ed.), A study of teaching: Multiple lenses, multiple views, Monograph no. 14 of the *Journal for Research in Mathematics Education* (pp. 13–44). Reston, VA: National Council of Teachers of Mathematics.

Carraher, D., & Nemirovsky, R. (Eds.) (2005). *Medium and meaning: Video papers in mathematics education research* (Neil A. Pateman, Series Ed.). Monograph no. 13 of the *Journal for Research in Mathematics Education*). Reston, VA: National Council of Teachers of Mathematics.

Engle, R. A., & Greeno, J. G. (1994) Managing disagreement in intellectual conversations: Coordinating interpersonal and conceptual concerns in the collaborative construction of mathematical explanations. In *Proceedings of the Sixteenth Annual Conference of the Cognitive Science Society.* Hillsdale, NJ: Erlbaum.

Hiebert, J., Gallimore, R., & Stigler, J. (2003). The new heroes of teaching. *Education Week, 23*(1), 56, 42. Retrieved December 13, 2005, from www.edweek/org/ew/ewstory.cfm?slug=10hiebert.h23

Horn, I. (2007). Accountable argumentation as a participation structure to support learning through disagreement. In Alan H. Schoenfeld (Ed.), *A study of teaching: Multiple lenses, multiple views,* Monograph no. 14 of the *Journal for Research in Mathematics Education* (pp. 97–126). Reston, VA: National Council of Teachers of Mathematics.

Hufferd-Ackles, K., Fuson, K., & Sherin, M. G. (2004). Describing levels and components of a math-talk community. *Journal for Research in Mathematics Education, 35* (2), 81–116.

Lewis, J. (2007). "Through the Looking Glass": A study of teaching. In Alan H. Schoenfeld (Ed.), *A study of teaching: Multiple lenses, multiple views,* Monograph no. 14 of the *Journal for Research in Mathematics Education* (pp. 1–12). Reston, VA: National Council of Teachers of Mathematics.

Posner, T. (2007). Looking at equity in mathematics classrooms. In Alan H. Schoenfeld (Ed.), A study of teaching: Multiple lenses, multiple views, Monograph no. 14 of the *Journal for Research in Mathematics Education* (pp. 127–172). Reston, VA: National Council of Teachers of Mathematics.

Schoenfeld, A. H. (1998). Toward a theory of teaching in context. *Issues in Education, 4*(1), 1–94.

Schoenfeld, A. H. (1999). Models of the teaching process. *Journal of Mathematical Behavior, 18*(3), 243–261.

Schoenfeld, A. H. (2007). On modeling teachers' in-the-moment decision-making. In Alan H. Schoenfeld (Ed.), *A study of teaching: Multiple lenses, multiple views,* Monograph no. 14 of the *Journal for Research in Mathematics Education* (pp. 45–96). Reston, VA: National Council of Teachers of Mathematics.

Sfard, A., & McClain, K. (2002). Analyzing tools: Perspectives on the role of designed artifacts in mathematics learning [Special issue]. *Journal of the Learning Sciences, 11*(2 & 3), 153–161.

Sherin, M. G. (2002). A balancing act: Developing a discourse community in a mathematics class-room. *Journal of Mathematics Teacher Education, 5,* 205–233.

Silver, E. A., & Smith, M. S. (1996). Building discourse communities in mathematics classrooms: A worthwhile but challenging journey: In P. C. Elliott (Ed.), *Communication in mathematics, K–12 and beyond: 1996 yearbook* (pp. 20–28). Reston, VA: National Council of Teachers of Mathematics.

Wood, T. (1999). Creating a context for argument in mathematics class. *Journal for Research in Mathematics Education, 30,* 171–191.

Appendix 1

Transcript of Lesson Segment

Deborah Ball's Third-Grade Class
Friday, January 19, 1990
Spartan Village School, East Lansing, Michigan

Line no.	Speaker	Transcript	Time
1	Ball:	[A]¹ Okay. A few delays, but I think we're ready to start now. [B] I'd like to open the discussion today with um—I have a few questions about the meeting yesterday that I'd like to ask. [C] So, to begin with, I would just like everybody to put pens down, there's nothing to take notes about or do right now. [D] But I'd like you to be thinking back to yesterday and to the meeting that we had on even and odd numbers and zero. [E] And I have a few questions. First—my first question is, I'd just like to hear some comments about what you thought about the meeting, what you noticed about the meeting, what you learned at the meeting, just what kinds of comments you have about yesterday's meeting? [F] And could you listen to one another's comments, so that we can um, benefit from what other people say? [G] See what y— what you think about other people's comments? Sheena, do you want to start?	12:59:15
2	Sheena:	I—I—I liked it because, well, I like talking to other classes and, and when you talk to other classes sometimes it helps.	1:00:06

---

¹ For ease of reference in the narrative, long statements made by Ball are broken into segments labeled [A], [B], [C], ….

Line			
no.	*Speaker*	*Transcript*	*Time*
3	Ball:	In what way?	1:00:21
4	Sheena:	It helps you to understand a little bit more.	
5	Ball:	Was there an example of something yesterday that you understood a little bit more during the meeting?	1:00:26
6	Sheena:	Well, I didn't think that zero was—zero, um—even or odd until yesterday they said that it could be even because of the ones on each side is odd, so that couldn't be odd. So that helped me understand it.	
7	Ball:	Hmm. So y— So you thought about something that came up in the meeting that you hadn't thought about before? Okay.	
8	Sheena:	(*nods*)	
9	Ball:	Other people's comments? Sean?	
10	Sean:	Um, I—I—I just want to say something to Sheena, when sh— what she said about um that, that one, um—zero has to be an odd, an even number bec— I disagree because, um, because what what two things can you put together to make it?	1:00:55
11	Sheena:	Could you repeat what you said, please?	1:01:13
12	Ball:	(*speaks to Betsy and asks her to listen to this*)	
13	Sean:	Okay, um, I disagree with you because, um, if it was an even number, how—what two things could make it?	
14	Sheena:	Well, I could show you it. (*Moves toward the chalkboard and points to the number line above the chalkboard.*) Um, I forgot what his name was—but yesterday he said that this one (*points to the 1 on the number line*) and each—this one is odd and this one (*points to the –1 on the number line*) is odd, so this one has to be even.	1:01:25
15	Sean:	But, that doesn't mean it always is even.	
16	Sheena:	It *could* be even.	
17	Sean:	It *could* be, but . . .	
18	Sheena:	I'm not saying that is *has* to be even. I meant that it could be.	
19	Sean:	You said it was.	

Line no.	Speaker	Transcript	Time
20	Ball:	[A] Before we take this up again, I underst— I—I understand that this is still a problem and that we didn't a— we didn't settle it, we're probably not going to settle it. [B] Um, there's a lot of disagreement about this issue, right? [C] And you saw that the fourth graders who have been thinking about this for a long time also disagree about it, don't they? [D] I'm still kind of interested um, in hearing some more comments about the meeting *itself*. [E] Sheena commented that it was good to have the two classes together because she heard an idea that she hadn't thought about and it made her think about and even revise her own idea when she was in the meeting yesterday. [F] What other comments do other people have about the meeting and what happened yesterday? [G] Mei, do you have a comment?	1:01:50
21	Mei:	Um, I h— I thought that zero was always going to be a even number, but from the meeting I sort of got mixed up because I heard other ideas I agree with and now I don't know which one I should agree with.	1:02:34
22	Ball:	Um-hm. So what are you going to do about that?	
23	Mei:	Um, I'm going to listen more to the discussion and find out.	
24	Ball:	Other people? Nathan?	1:02:56
25	Nathan:	Um, first I said that um, zero was even but then I guess I revised so that zero, I think, is special because um, I—um, even numbers, like they they *make* even numbers; like two, um, two makes four, and four is an even number; and four makes eight; eight is an even number; and um, like that. And, and go on like that and like one plus one and go on adding the same numbers with the same numbers. And so I, I think zero's special.	

Line no.	Speaker	Transcript	Time
26	Ball:	**[A]** Can I ask you a question about what you just said? |**[B]** And then I'll ask people for more comments about the meeting. **[C]** Were you saying that when you put even numbers together, you get another even number—	1:03:41
27	Nathan:	Yeah.	
28	Ball:	—or were you saying that all even numbers are made up of even numbers?	
29	Nathan:	Yes, they are. [This comment is very hard to make out. Significant dispute has occurred over whether Nathan said "Yes, they are" or "No, they're not."]	
30	Ball:	Betsy, you said something like that yesterday, too.	1:04:00
31	Betsy:	What?	
32	Ball:	Were you—were you not listening to this just now?	
33	Betsy:	No.	
34	Ball:	Nathan said a minute ago that when you put even numbers together you get an even number,	1:04:06
35	Betsy:	Mm-hm.	
36	Ball:	But he also said, I think, that all even numbers are made up of other even numbers.	
37	Mei:	I disagree.	
38	Sheena:	(*says something to Mei*)	
39	Ball:	Two even numbers just the same.	
40	Nathan:	Unh-uh.	
41	Ball:	The same even number?	
42	Nathan:	Yeah, like four.	
43	Ball:	**[A]** Like eight is four plus four? **[B]** Are all the even numbers—can you do that with all the even numbers? That they'd be made up of two identical even numbers?	
44	Sean:	Not—not—not—	1:04:31
45	Betsy:	(*looking toward Nathan*) You can't. Like six. Six is two, two, ... Six you can't get two.	

Line no.	Speaker	Transcript	Time
46	Sean:	Six is two *odd* numbers to make an even, to make an even number.	
47	Mei:	Three and three—	
48	Betsy:	(*still looking toward Nathan*) You need three twos to make six. You can't put a four and a four or a . . .	
49	Sean:	Three twos???	
50	Betsy:	(*looking toward Nathan*) Three's—Three is odd.	
51	Sean:	Or, um—	
52	Nathan:	I know that, but um, um I'm talking about like two plus two is four, and four plus four is eight and I just skipped the six so I just added the ones that, that add. Like, the two plus two is four, and four is an even number and I'm just talking about the things that um, like—	
53	Sean:	Six can be an odd number.	
54	Nathan:	what I just said—the, um, like two is plus two is four and four plus four is eight and—	
55	Betsy:	So what you're doing is you're going by twos and then what two equals from then you go from—all the way up.	
56	Nathan:	Yeah, I'm not going by every single number. Like,	
57	Betsy:	Okay.	
58	Nathan:	two, four, six, eight.	
59	Ball:	[A] More comments about the meeting? [B] I'd really like to hear from as many people as possible what comments you had or reactions you had to being in that meeting yesterday. [C] Sean?	1:05:31
60	Sean:	Um, I don't have anything about the meeting yesterday, but I was just thinking about six, that it's a . . . I'm just thinking. I'm just thinking it can be an odd number, too, 'cause there could be two, four, six, and two, three twos, that'd make six . . .	
61	Ball:	Uh-huh . . .	

Line			
no.	*Speaker*	*Transcript*	*Time*
62	Sean:	And two *threes,* that it could be an odd and an *even* number. Both. *Three* things to make it, and there could be *two* things to make it.	
63	Ball:	And the two things that you put together to make it were odd, right? Three and three are each *odd?*	
64	Sean:	Uh huh, and the other, the twos were even.	
65	Ball:	[A] So you're kind of—I think Nathan said then that he wasn't talking about *every* even number, right, Nathan? [B] Were you saying that? [C] Some of the even numbers, like six, are made up of two odds, like you just suggested.	
66	Nathan:	Uh-uh (agreeing with the teacher).	
67	Ball:	Other people's comments?	

Student Demographics

Name	Gender	Race	Country	English proficiency	How long at this school
Lindiwe	M	African American	U.S.A./ South Africa	Fluent	2 weeks
Nathan	M	White	Ethiopia	Fluent	3 years
Betsy	F	White	Canada	Native speaker	4 months
Daniel	M	Asian	Indonesia	Developing	3 years
Jeannie	F	White	U.S.A.	Native speaker	3 years
Keith	M	African American	U.S.A.	Native speaker	2 weeks
Tempe	M	African Black	Kenya	Fluent	3 years
Mei	F	Asian	Taiwan	Fluent	2 years
Lucy	F	White	U.S.A.	Native speaker	3 years
Maria	F	Latina	Nicaragua	Beginning	4 months
Mark	M	White	U.S.A.	Native speaker	2 years
Ofala	F	African Black	Nigeria	Fair	3 years
Devin	M	White	Nepal	Beginning	9 months
Riba	F	White	Egypt	Good	3 years
Harooun	M	Asian	Indonesia	Developing	16 months
Sean	M	White	U.S.A.	Native speaker	2 years
Sheena	F	African American	U.S.A.	Native speaker	4 months
Tory	F	White	U.S.A.	Native speaker	2 weeks
Cassandra	F	African American	U.S.A.	Native speaker	16 months